Bluetooth Low Energy

Bluetooth Low Energy

The Developer's Handbook

Robin Heydon

PRENTICE HALL

Upper Saddle River, NJ • Boston • Indianapolis • San Francisco
New York • Toronto • Montreal • London • Munich • Paris • Madrid
Capetown • Sydney • Tokyo • Singapore • Mexico City

Many of the designations used by manufacturers and sellers to distinguish their products are claimed as trademarks. Where those designations appear in this book, and the publisher was aware of a trademark claim, the designations have been printed with initial capital letters or in all capitals.

The author and publisher have taken care in the preparation of this book, but make no expressed or implied warranty of any kind and assume no responsibility for errors or omissions. No liability is assumed for incidental or consequential damages in connection with or arising out of the use of the information or programs contained herein.

The publisher offers excellent discounts on this book when ordered in quantity for bulk purchases or special sales, which may include electronic versions and/or custom covers and content particular to your business, training goals, marketing focus, and branding interests. For more information, please contact:

> U.S. Corporate and Government Sales
> (800) 382-3419
> corpsales@pearsontechgroup.com

For sales outside the United States please contact:

> International Sales
> international@pearson.com

Visit us on the Web: informit.com/ph

Cataloging-in-Publication Data is on file with the Library of Congress.

Copyright © 2013 Pearson Education, Inc.

All rights reserved. Printed in the United States of America. This publication is protected by copyright, and permission must be obtained from the publisher prior to any prohibited reproduction, storage in a retrieval system, or transmission in any form or by any means, electronic, mechanical, photocopying, recording, or likewise. To obtain permission to use material from this work, please submit a written request to Pearson Education, Inc., Permissions Department, One Lake Street, Upper Saddle River, New Jersey 07458, or you may fax your request to (201) 236-3290.

ISBN-13: 978-0-13-288836-3
ISBN-10:   0-13-288836-X

Text printed in the United States on recycled paper at RR Donnelley in Crawfordsville, Indiana.
First printing, October 2012

Executive Editor
Bernard Goodwin

Managing Editor
John Fuller

Project Editor
Elizabeth Ryan

Copy Editor
Bob Russell

Indexer
Jack Lewis

Proofreader
Christine Clark

Cover Designer
Gary Adair

Compositor
LaurelTech

This book is dedicated to Katherine.

—Robin xxx

Contents

Preface	xvii
Acknowledgments	xix
About the Author	xxi

Part I	**Overview**	**1**
Chapter 1	**What Is Bluetooth Low Energy?**	**3**
1.1	Device Types	6
1.2	Design Goals	7
1.3	Terminology	9
Chapter 2	**Basic Concepts**	**11**
2.1	Button-Cell Batteries	11
2.2	Time Is Energy	12
2.3	Memory Is Expensive	13
2.4	Asymmetric Design	14
2.5	Design For Success	15
2.6	Everything Has State	16
2.7	Client-Server Architecture	17
2.8	Modular Architecture	18
2.9	One Billion Is a Small Number	19
2.10	Connectionless Model	19
2.11	Paradigms	20
	2.11.1 Client-Server Architecture	20
	2.11.2 Service-Oriented Architecture	21

Chapter 3   Architecture — 27
3.1   Controller — 27
    3.1.1   Physical Layer — 28
    3.1.2   Direct Test Mode — 29
    3.1.3   Link Layer — 30
    3.1.4   The Host/Controller Interface — 31
3.2   The Host — 32
    3.2.1   Logical Link Control and Adaptation Protocol — 32
    3.2.2   The Security Manager Protocol — 33
    3.2.3   The Attribute Protocol — 33
    3.2.4   The Generic Attribute Profile — 34
    3.2.5   The Generic Access Profile — 36
3.3   The Application Layer — 36
    3.3.1   Characteristics — 36
    3.3.2   Services — 37
    3.3.3   Profiles — 37
3.4   Stack Splits — 38
    3.4.1   Single-Chip Solutions — 38
    3.4.2   Two-Chip Solutions — 39
    3.4.3   Three-Chip Solutions — 40

Chapter 4   New Usage Models — 41
4.1   Presence Detection — 41
4.2   Broadcasting Data — 42
4.3   Connectionless Model — 43
4.4   Gateways — 44

Part II   Controller — 47

Chapter 5   The Physical Layer — 49
5.1   Background — 49
5.2   Analog Modulation — 49
5.3   Digital Modulation — 51
5.4   Frequency Band — 54
5.5   Modulation — 54
5.6   Radio Channels — 55
5.7   Transmit Power — 56
5.8   Tolerance — 57

5.9	Receiver Sensitivity		57
5.10	Range		58

Chapter 6 Direct Test Mode — 61

6.1	Background		61
6.2	Transceiver Testing		62
	6.2.1	Test Packet Format	63
	6.2.2	Transmitter Tests	63
	6.2.3	Receiver Tests	64
6.3	Hardware Interface		65
	6.3.1	UART	65
	6.3.2	Commands and Events	65
6.4	Direct Testing by Using HCI		67

Chapter 7 The Link Layer — 69

7.1	The Link Layer State Machine		69
	7.1.1	The Standby State	70
	7.1.2	The Advertising State	71
	7.1.3	The Scanning State	72
	7.1.4	The Initiating State	73
	7.1.5	The Connection State	73
	7.1.6	Multiple State Machines	74
7.2	Packets		76
	7.2.1	Advertising and Data Packets	76
	7.2.2	Whitening	77
7.3	Packet Structure		79
	7.3.1	Bit Order and Bytes	79
	7.3.2	The Preamble	79
	7.3.3	Access Address	80
	7.3.4	Header	81
	7.3.5	Length	82
	7.3.6	Payload	83
	7.3.7	Cyclic Redundancy Check	84
7.4	Channels		84
	7.4.1	Frequency Hopping	87
	7.4.2	Adaptive Frequency Hopping	88
7.5	Finding Devices		90
	7.5.1	General Advertising	91
	7.5.2	Direct Advertising	91

	7.5.3	Nonconnectable Advertising	92
	7.5.4	Discoverable Advertising	92
7.6	Broadcasting	92
7.7	Creating Connections	93
	7.7.1	Access Address	95
	7.7.2	CRC Initialization	95
	7.7.3	Transmit Window	95
	7.7.4	Connection Events	96
	7.7.5	Channel Map	97
	7.7.6	Sleep Clock Accuracy	98
7.8	Sending Data	98
	7.8.1	Data Header	99
	7.8.2	Logical Link Identifier	100
	7.8.3	Sequence Numbers	101
	7.8.4	Acknowledgement	101
	7.8.5	More Data	101
	7.8.6	Examples of the Use of Sequence Numbers and More Data	101
7.9	Encryption	104
	7.9.1	AES	105
	7.9.2	Encrypting Payload Data	106
	7.9.3	Message Integrity Check	107
7.10	Managing Connections	109
	7.10.1	Connection Parameter Update	109
	7.10.2	Adaptive Frequency Hopping	111
	7.10.3	Starting Encryption	112
	7.10.4	Restarting Encryption	115
	7.10.5	Version Exchange	117
	7.10.6	Feature Exchange	118
	7.10.7	Terminating Connections	118
7.11	Robustness	120
	7.11.1	Adaptive Frequency Hopping	120
	7.11.2	Strong CRCs	122
7.12	Optimizations for Low Power	123
	7.12.1	Short Packets	124
	7.12.2	High Bit Rate	125
	7.12.3	Low Overhead	126
	7.12.4	Acknowledgement Scheme	127
	7.12.5	Single-Channel Connection Events	127

	7.12.6	Subrating Connection Events	128
	7.12.7	Offline Encryption	130

Chapter 8 The Host/Controller Interface — 131

- 8.1 Introduction — 131
- 8.2 Physical Interfaces — 131
  - 8.2.1 UART — 132
  - 8.2.2 3-Wire UART — 132
  - 8.2.3 USB — 134
  - 8.2.4 SDIO — 134
- 8.3 Logical Interface — 135
  - 8.3.1 HCI Channels — 135
  - 8.3.2 Command Packets — 135
  - 8.3.3 Event Packets — 137
  - 8.3.4 Data Packets — 138
  - 8.3.5 Command Flow Control — 139
  - 8.3.6 Data Flow Control — 140
- 8.4 Controller Setup — 140
  - 8.4.1 Reset the Controller to a Known State — 141
  - 8.4.2 Reading the Device Address — 141
  - 8.4.3 Set Event Masks — 142
  - 8.4.4 Read Buffer Sizes — 142
  - 8.4.5 Read Supported Features — 143
  - 8.4.6 Read Supported States — 144
  - 8.4.7 Random Numbers — 145
  - 8.4.8 Encrypting Data — 145
  - 8.4.9 Set Random Address — 146
  - 8.4.10 White Lists — 147
- 8.5 Broadcasting and Observing — 148
  - 8.5.1 Advertising — 148
  - 8.5.2 Passive Scanning — 150
  - 8.5.3 Active Scanning — 152
- 8.6 Initiating Connections — 153
  - 8.6.1 Initiating Connection to White List — 154
  - 8.6.2 Initiating a Connection to a Device — 156
  - 8.6.3 Canceling Initiating a Connection — 156
- 8.7 Connection Management — 158
  - 8.7.1 Connection Update — 158
  - 8.7.2 Channel Map Update — 159

8.7.3	Feature Exchange	160
8.7.4	Version Exchange	160
8.7.5	Starting Encryption	161
8.7.6	Restarting Encryption	163
8.7.7	Terminating a Connection	164

Part III Host   167

Chapter 9 Logical Link Control and Adaptation Protocol   169
9.1 Background   169
9.2 L2CAP Channels   171
9.3 The L2CAP Packet Structure   172
9.4 The LE Signaling Channel   173
    9.4.1 Command Reject   174
    9.4.2 Connection Parameter Update Request and Response   175

Chapter 10 Attributes   179
10.1 Background   179
    10.1.1 Protocol Proliferation Is Wrong   180
    10.1.2 Data, Data, Everywhere...   180
    10.1.3 Data and State   181
    10.1.4 Kinds of State   182
    10.1.5 State Machines   183
    10.1.6 Services and Profiles   185
10.2 Attributes   189
    10.2.1 Attribute   189
    10.2.2 The Attribute Handle   189
    10.2.3 Attribute Type   190
    10.2.4 Attribute Value   191
    10.2.5 Databases, Servers, and Clients   192
    10.2.6 Attribute Permissions   194
    10.2.7 Accessing Attributes   196
    10.2.8 Atomic Operations and Transactions   197
10.3 Grouping   199
10.4 Services   199
    10.4.1 Extending Services   201
    10.4.2 Reusing Another Service   203
    10.4.3 Combining Services   204
    10.4.4 Primary or Secondary   205

	10.4.5	Plug-and-Play Client Applications	207
	10.4.6	Service Declaration	208
	10.4.7	Including Services	209
10.5	Characteristics		210
	10.5.1	Characteristic Declaration	211
	10.5.2	Characteristic Value	213
	10.5.3	Descriptors	214
10.6	The Attribute Protocol		217
	10.6.1	Protocol Messages	219
	10.6.2	The Exchange MTU Request	221
	10.6.3	The Find Information Request	221
	10.6.4	The Find By Type Value Request	222
	10.6.5	The Read By Type Request	223
	10.6.6	The Read Request	224
	10.6.7	The Read Blob Request	224
	10.6.8	The Read Multiple Request	224
	10.6.9	The Read By Group Type Request	225
	10.6.10	The Write Request	225
	10.6.11	The Write Command	225
	10.6.12	The Signed Write Command	225
	10.6.13	The Prepare Write Request and Execute Write Request	226
	10.6.14	The Handle Value Notification	227
	10.6.15	The Handle Value Indication	228
	10.6.16	Error Response	228
10.7	The Generic Attribute Profile		231
	10.7.1	The Discovery Procedures	232
	10.7.2	The Discovering Services	232
	10.7.3	Characteristic Discovery	234
	10.7.4	Client-Initiated Procedures	235
	10.7.5	Server-Initiated Procedures	238
	10.7.6	Mapping ATT PDUs to GATT Procedures	239

Chapter 11   Security — **241**

11.1	Security Concepts		241
	11.1.1	Authentication	241
	11.1.2	Authorization	242
	11.1.3	Integrity	243
	11.1.4	Confidentiality	243
	11.1.5	Privacy	243

	11.1.6	Encryption Engine	244
	11.1.7	Shared Secrets	244
11.2	Pairing and Bonding		248
	11.2.1	Pairing	248
	11.2.2	Exchange of Pairing Information	248
	11.2.3	Authentication	250
	11.2.4	Key Distribution	251
	11.2.5	Bonding	252
11.3	Signing of Data		252

Chapter 12 The Generic Access Profile — **255**

12.1	Background		255
	12.1.1	Initial Discovery	256
	12.1.2	Establishing the Initial Connection	258
	12.1.3	Service Characterization	258
	12.1.4	Long-Term Relationships	259
	12.1.5	Reconnections	260
	12.1.6	Private Addresses	260
12.2	GAP Roles		261
12.3	Modes and Procedures		262
	12.3.1	Broadcast Mode and Observation Procedure	263
	12.3.2	Discoverability	263
	12.3.3	Connectability	266
	12.3.4	Bonding	270
12.4	Security Modes		270
	12.4.1	Security Modes	271
12.5	Advertising Data		273
	12.5.1	Flags	273
	12.5.2	Service	274
	12.5.3	Local Name	275
	12.5.4	TX Power Level	275
	12.5.5	Slave Connection Interval Range	275
	12.5.6	Service Solicitation	275
	12.5.7	Service Data	276
	12.5.8	Manufacturer-Specific Data	276
12.6	GAP Service		276
	12.6.1	The Device Name Characteristic	276
	12.6.2	The Appearance Characteristic	276

Contents   xv

    12.6.3   The Peripheral Privacy Flag   277
    12.6.4   Reconnection Address   278
    12.6.5   Peripheral Preferred Connection Parameters   278

Part IV   Application   281

Chapter 13   Central   283
13.1   Background   283
13.2   Discovering Devices   283
13.3   Connecting to Devices   285
13.4   What Does This Device Do?   286
13.5   Generic Clients   287
13.6   Interacting with Services   288
    13.6.1   Readable Characteristics   288
    13.6.2   Control Points   289
    13.6.3   State Machines   290
    13.6.4   Notifications and Indications   291
13.7   Bonding   292
13.8   Changed Services   293
13.9   Implementing Profiles   294
    13.9.1   Defining a Profile   294
    13.9.2   Finding Services   295
    13.9.3   Finding Characteristics   296
    13.9.4   Using Characteristics   296
    13.9.5   Profile Security   296

Chapter 14   Peripherals   299
14.1   Background   299
14.2   Broadcast Only   299
14.3   Being Discoverable   300
14.4   Being Connectable   301
14.5   Exposing Services   301
14.6   Characteristics   302
14.7   Security Matters   303
14.8   Optimizing for Low Power   303
    14.8.1   Discoverable Advertising   305
    14.8.2   Bonding   306
    14.8.3   Connectable Advertising   306

	14.8.4	Directed Advertising	307
	14.8.5	Connected	307
	14.8.6	Stay Connected or Disconnect	309
14.9	Optimizing Attributes		311

Chapter 15 Testing and Qualification   313

15.1 Starting a Project   313
15.2 Selecting Features   316
15.3 Consistency Check   316
15.4 Generating a Test Plan   317
15.5 Creating a Compliance Folder   317
15.6 Qualification Testing   318
15.7 Qualify Your Design   319
15.8 Declaring Compliance   320
15.9 Listing   321
15.10 Combining Components   321

Index   **323**

Preface

Sometimes, once in a lifetime, a new technology comes along that changes the world; for example, AM radio, television, and wireless Internet. Bluetooth low energy is at the cusp of the next revolution in wireless technology: a technology that can be embedded in products because it uses so little power that it can be designed around a small battery that lasts for years.

This book explains how this technology came about, why it was designed the way it has been designed, and how it works. It is written by one of the leading experts on Bluetooth low energy, Robin Heydon, who has been involved in creating the specifications, interoperability testing, and training.

This book is for anyone who is thinking about developing a product that incorporates Bluetooth low energy, whether you are an engineer, an application developer, a designer, or you're in marketing.

For engineers, the book covers the details of how the complete system works, from the physical radio waves up to the discovery of, connection with, and interface provided by that device.

For application developers, this book provides an understanding of the constraints imposed by Bluetooth low energy on applications. It also presents a thorough description of the design goals and implementation of these requirements.

For designers, the information contained herein will allow you to appreciate the particular problems with designing Bluetooth low energy wireless products, from how the product might need to work and how big a battery might be required to implement your ideas.

For everyone else, the book provides the background of why Bluetooth low energy was designed, the design goals it tried to achieve, and how you can take something that radically changes the way you can think of wireless technology and implement it in everything else.

The book is split into four parts:

Part I provides an overview of the technology, the basic concepts that guided the development of Bluetooth low energy, the architecture of the system from the radio through the various protocol layers up to the application layers, and finally, the new usage models that this new technology enables.

The second part goes into detail on how the radio chip—called a controller—functions. This is the silicon chip that product designers need to incorporate into their end products. This part also covers the radio, Direct Test Mode, and the Link Layer. In addition, it shows how to interact with the controller from the upper-layer stack, called a host.

Part III goes into detail of how the host (the software stack) works. It covers the concepts and details behind the main protocol used to expose attributes of a device. It also covers the security models and how to make connections and bonds, or associate, two devices with one another.

In Part IV, you wrap up all the details by looking at the design considerations that a product or application developer needs to consider. It starts by looking at the issues involving central devices. Next, it looks at issues related to peripheral devices. Finally, it considers the entire problem surrounding testing and qualification, typically the final part of any product that will be taken to market.

If after reading the book you would like to learn more about Bluetooth low energy, there are a number of resources available. The specifications themselves are available on the Bluetooth SIG website at www.bluetooth.org. If you would like to find developer information about Bluetooth low energy, there is also a developer site available at developer.bluetooth.org that has detailed information about characteristics. The author also has a website at www.37channels.com, where you can view frequently asked questions raised by this book and Bluetooth low energy.

Acknowledgments

I would like to thank the following people for their invaluable help in making this book possible. Katherine Heydon, for reading the whole book cover to cover many times and providing constructive criticism on the contents. Jennifer Bray for her encouragement to write the book in the first place and allowing me the time and space to undertake such a task. All the production team at Addison-Wesley, especially Bernard Goodwin, Elizabeth Ryan, Michelle Housley and Gary Adair; my copy editor, Bob Russell; and all the others in the background who made this book happen. Nick Hunn for the many times spent discussing the best way to communicate the ideas behind the low energy technology. Zoë Hunn for the fantastic artwork on the front cover. Andy Glass for constantly asking (nagging?) about when the book would be done and providing excellent review comments. Steve Wenham, who suffered my constant ideas about how low energy could be made better. British Airways, for almost always giving me a front row bulkhead seat and allowing me to use my Bluetooth keyboard and mouse on the many long-haul flights. This book was probably written at an average height of 30,000 feet. For the Bluetooth SIG community in general, for the many questions that they asked at All Hands Meetings, UnPlugFests, and all the various working group meetings: these questions helped determine what were the hardest concepts to explain, and therefore the basic structure and contents of this book.

About the Author

Robin Heydon was educated as a software engineer, graduating with a degree in Computer Science from the University of Manchester, UK. He was employed in the computer entertainment industry for a decade working on networked flight simulators. He then moved into wireless communications in 2000, working for what was then a small company called CSR. There he moved from being a firmware engineer to working as a full-time standards architect. In this work, Robin has worked on fixing and improving all versions of the Bluetooth specification. In early 2007, Robin started working on a project called Wibree, which later became the Bluetooth low energy specification. He cochaired the group, and drove through the specification to publication, and was recognized by the Bluetooth SIG as an inductee to the Bluetooth SIG Hall of Fame in 2010.

Part I

Overview

Chapter 1, What Is Bluetooth Low Energy?, introduces Bluetooth low energy, and discusses its design goals.

Chapter 2, Basic Concepts, discusses the foundations upon which the low energy architecture was designed.

Chapter 3, Architecture, introduces the main system architecture for low energy, from the controller, through the host, and up to the applications.

Chapter 4, New Usage Models, describes the new usage models that the low energy technology enables.

Chapter 1

What Is Bluetooth Low Energy?

> *If I have seen a little further,*
> *it is by standing on the shoulders of Giants.*
> —Isaac Newton

Bluetooth low energy is a brand new technology that has been designed as both a complementary technology to classic Bluetooth as well as the lowest possible power wireless technology that can be designed and built. Although it uses the Bluetooth brand and borrows a lot of technology from its parent, Bluetooth low energy should be considered a different technology, addressing different design goals and different market segments.

Classic Bluetooth was designed to unite the separate worlds of computing and communications, linking cell phones to laptops. However its killer application has proved to be as an audio link from the cell phone to a headset placed on or around the ear. As the technology matured, more and more use cases were added, including stereo music streaming, phone book downloads from the phone to your car, wireless printing, and file transfer. Each of these new use cases required more bandwidth, and therefore, faster and faster radios have been constantly added to the Bluetooth ecosystem over time. Bluetooth started with Basic Rate (BR) with a maximum Physical Layer data rate of 1 megabit per second (Mbps). Enhanced Data Rate (EDR) was added in version 2.0 of Bluetooth to increase the Physical Layer data rates to 3Mbps; an Alternate MAC[1] PHY[2] (AMP) was added in version 3.0 of Bluetooth that used IEEE[3] 802.11 to deliver Physical Layer data rates of up to hundreds of megabits per second.

Bluetooth low energy takes a completely different direction. Instead of just increasing the data rates available, it has been optimized for ultra-low power consumption. This means that you probably won't get high data rates, or even want to keep a connection up for many hours or days. This is an interesting move,

---

1. MAC stands for Medium Access Control. How a transceiver uses a Physical Layer to communicate with other transceivers.
2. PHY stands for Physical Layer.
3. IEEE stands for the Institute of Electrical and Electronics Engineers.

Table 1–1 Speeds Almost Always Increase

Modems	Ethernet
V.21: 0.3kbps	802.3i: 10Mbps
V.22: 1.2kbps	802.3u: 100Mbps
V.32: 9.6kbps	802.3ab: 1000Mbps
V.34: 28.8kbps	802.3an: 10000Mbps
Wi-Fi	Bluetooth
802.11: 2Mbps	v1.1: 1Mbps
802.11b: 11Mbps	v2.0: 3Mbps
802.11g: 54Mbps	v3.0: 54Mbps
802.11n: 135Mbps	v4.0: 0.3Mbps

as most wired and wireless communications technologies constantly increase speeds, as illustrated in Table 1–1.

This different direction has been achieved through the understanding that classic Bluetooth technology cannot achieve the low power requirements required for devices powered by button-cell batteries. However, to fully understand the requirements around low power, another consideration must be taken. Bluetooth low energy is also designed to be deployed in extremely high volumes, in devices that today do not have any wireless technology. One method to achieve very high volumes is to be extremely low cost. For example, Radio frequency identification (RFID) tags can be deployed in very high volumes because they are very low cost, ultimately because they work by scavenging power delivered by a more expensive scanner.

Therefore, it is crucial to also look at the Bluetooth low energy system design from the requirements of low cost. Three key elements within this design point to very low cost:

1. ISM Band

    The 2.4GHz ISM band is a terrible place to design and use a wireless technology. It has poor propagation characteristics, with the radio energy readily being absorbed by everything, but especially by water; consider that the human body is made up primarily of water. These rather significant downsides are made up by the fact that the radio spectrum is available worldwide and there are no license requirements. Of course, this Free Rent sign means that other technologies are also going to use this space, including most Wi-Fi radios. But the lack of licensing doesn't mean that anything goes. There are still plenty of rules, mainly related to limiting the power output of devices that use the spectrum, limiting the range. However, these limitations are still more attractive than paying heavily for licensed spectrum. Therefore, choosing to use the ISM band lowers the cost.

2. IP License

   When the Wibree technology was mature enough to be merged into an established wireless standards group, Nokia could have taken the technology to any such group. For example, it could have taken it to the Wi-Fi Alliance, which also standardizes technology in the same 2.4GHz ISM band. But they chose the Bluetooth Special Interest Group (SIG) because of the excellent reputation and licensing policy that this organization has. These policies basically mean that the patent licensing costs are significantly reduced for a Bluetooth device when compared with a technology developed in another SIG or association that has a FRAND[4] policy. Because Bluetooth has a very low license costs, the cost per device is also significantly reduced.

3. Low Power

   The best way to design a low-cost device is to reduce the materials required to make such a device—materials such as batteries. The larger the battery, the larger the battery casing needs to be, again increasing the costs. Replacing a battery costs money, not just for a consumer who needs to purchase another battery, but replacement also includes the opportunity costs of not having that device available. If this device is maintained by a third party, perhaps because it is part of a managed home alarm system, there are additional labor costs to change this battery. Therefore, designing the technology around low power consumption also reduces the costs. As a thought experiment, how would things be different if a megawatt battery were available for a single penny?

   Many devices could accommodate a larger battery. A keyboard or mouse can easily take AA batteries, yet the manufacturers want to use AAA batteries not because they are smaller, but because their use reduces the bill of materials and therefore the cost of the device.

Therefore, the fundamental design for low energy is to work with button-cell batteries—the smallest, cheapest, and most readily available type of battery available. This means that you cannot achieve high data rates or make low energy work for use cases that require large data transfers or the streaming of data. This single point

---

4. FRAND stands for Fair, Reasonable, and Non-Discriminatory. This means that if you license your technology, you must do it at a fair price, on the same terms for everybody, regardless of who the licensee is.

is probably the most important difference between classic and low-energy variants of Bluetooth. This is discussed further in the next section.

1.1 Device Types

Bluetooth low energy makes it possible to build two types of devices: dual-mode and single-mode devices. A dual-mode device is a Bluetooth device that has support for both Bluetooth classic as well as Bluetooth low energy. A single-mode device is a Bluetooth device that only supports Bluetooth low energy. There is a third type of device, which is a Bluetooth classic-only device.

Because it supports Bluetooth classic, a dual-mode device can talk with the billions of existing Bluetooth devices. Dual-mode devices are new. They require new hardware and firmware in the controller and software in the host. It is therefore not possible to take an existing Bluetooth classic controller or host and upgrade it to support low energy. However, most dual-mode controllers are simple replacement parts for existing Bluetooth classic controllers. This allows designers of cell phones, computers, and other device to replace their existing Bluetooth classic controllers with dual-mode controllers very quickly.

Because it does not support Bluetooth classic, a Bluetooth low energy single-mode device cannot talk with the existing Bluetooth devices, but it can still talk with other single-mode devices as well as dual-mode devices. These new single-mode devices are highly optimized for ultra-low power consumption, being designed to go into components that are powered by button-cell batteries. Single-mode devices will also not be able to be used in most of the use cases for which Bluetooth classic is used today because single-mode Bluetooth low energy does not support audio for headsets and stereo music or high data rates for file transfers.

Table 1–2 shows what device types can talk with other devices types and what Bluetooth radio technology would be used when they connect. Single-mode devices will talk with other single-mode devices using low energy. Single-mode devices will also talk with dual-mode devices using low energy. Dual-mode devices will talk with other dual-mode devices or classic devices using BR/EDR. A single-mode device cannot talk with a classic device.

Table 1–2 Single-Mode, Dual-Mode, and Classic Compatibility

	Single-Mode	Dual-Mode	Classic
Single-Mode	LE	LE	none
Dual-Mode	LE	Classic	Classic
Classic	none	Classic	Classic

1.2 Design Goals

When reviewing any technology, the first question to be asked is how did the designers optimize this technology? Most technologies have one or two things that they are very good at, and many things that they are not. By determining what these one or two things are, a greater understanding of that technology can be achieved.

With Bluetooth low energy, this is very simple. It was designed for ultra-low power consumption. The unique structure of the Bluetooth SIG is that the organization creates and controls everything from the Physical Layer up to the application. The SIG does this in a cooperative and open but commercially driven standards model, and over more than ten years, it has optimized the process of creating wireless specifications that not only work at the point of release but are also interoperable, robust, and of extremely high quality.

When the low energy work started, the goal was to create the lowest-power short-range wireless technology possible. To do this, each layer of the architecture has been optimized to reduce the power consumption required to perform a given task. For example, the Physical Layer's relaxation of the radio parameters, when compared with a Bluetooth classic radio, means that the radio can use less power when transmitting or receiving data. The link layer is optimized for very rapid reconnections and the efficient broadcast of data so that connections may not even be needed. The protocols in the host are optimized to reduce the time required once a link layer connection has been made until the application data can be sent. All of this is possible only when all parts of the system are designed at the same time by the same group of people.

The design goals for the original Bluetooth radio have not been forgotten. These include the following:

- Worldwide operation
- Low cost
- Robust
- Short range
- Low power

For global operation, a wireless band that is available worldwide is required. There is only one available band that can be implemented using low-cost and high-volume manufacturing technology today: the 2.45GHz band. This is available because it is of no interest to astronomers, cell phone operators, or other commercial interests. Unfortunately, just like everything that is free, everybody wants to be part of it, causing congestion. Other wireless bands are available, for example, the 60GHz ISM band, but this is not practical from a low-cost point of view, or the 800/900MHz bands that have different frequencies and rules depending on where you are on the planet.

The design goal of low cost is interesting because it implies that the system should be kept as small and efficient as possible. Although it could be possible, for example, to add scatter net support or full-mesh networking into Bluetooth low energy, this would increase the cost because more memory and processing power would be required to maintain this network. The system has therefore been optimized for low cost above interesting research-based networking topologies.

The 2.45GHz band that Bluetooth low energy uses is already very crowded. Just taking into account standards-based technologies, it includes Bluetooth classic, Bluetooth low energy, IEEE 802.11, IEEE 802.11b, IEEE 802.11g, IEEE 802.11n, and IEEE 802.15.4. In addition, a number of proprietary radios are also using the band, including X10 video repeaters, wireless alarms, keyboards, and mice. A number of devices also emit noise in the band, such as street lights and microwave ovens.

It is therefore almost impossible to design a radio that will work at all times with all possible interferers, unless it uses *adaptive frequency hopping*, as pioneered by Bluetooth classic. Adaptive frequency hopping helps by not only detecting sources of interference quickly but also by adaptively avoiding them in the future. It also quickly recovers from the inevitable dropped packets caused by interference from other radios. It is this robustness that is absolutely key to the success of any wireless technology in the most congested radio spectrum available.

Robustness also covers the ability to detect and recover from bit errors caused by background noise. Most short-range wireless standards compromise by using a short cyclic redundancy check (CRC), although there are some that use very long checks. A good design will see compromise between the strength of the checks and the time taken to send this information.

Short range is actually a slight problem. If you want a low-power system, you must keep the transmitted power as low as possible to reduce the energy used to transmit the signal. Similarly, you must keep the receiver sensitivity fairly high to reduce the power required to pick up the radio signals of other devices from amongst the noise. What short range means in this context is really that it is not centered around a cellular base station system. Short range means that Bluetooth low energy should be a *personal area network*.

The original Bluetooth design goal of low power hasn't changed that much, except that the design goals for power consumption have been reduced by one or two orders of magnitude. Bluetooth classic had a design goal of a few days standby and a few hours talk time for a headset, whereas Bluetooth low energy has a design goal of a few years for a sensor measuring the temperature or measuring how far you've walked.

1.3 Terminology

Just like many high technology areas, the people working in Bluetooth low energy use their own language to describe the features and technology with the specification. This section enumerates each of the words that have special meaning and what they mean.

Adaptive Frequency Hopping (AFH) A technology whereby only a subset of frequencies is used. This allows devices to avoid the frequencies that other non-adaptive technologies are using (e.g., a Wi-Fi access point).

Architecture The design of the Bluetooth low energy is sometimes known as the Architecture.

Band See Radio Band.

Frequency Hopping The use of multiple frequencies to communicate between two devices. One frequency is used at a time, and each frequency is used in a defined sequence.

Layer A part of the system that fulfills a specific function. For example, the Physical Layer covers the operation of the radio. Each layer in a system is abstracted away from the layers above and below it. The Link Layer doesn't need to know all the details of how the radio functions; the Logical Link Control Layer and Adaptation Layer don't need to know all the details of how the Link Layer works. This abstraction is important to keep the complexity of the system at manageable levels.

Master A complex device that coordinates the activity of other devices within a piconet.

Piconet This is a contraction of the words pico and network. Pico is the SI[5] prefix for 10^{-12}. This is derived from the Italian *piccolo*, meaning small.[6] Therefore, a piconet is a very small network. A piconet has a single master device that coordinates the activity of all the other devices (slaves) in the piconet and one or more slaves.

Radio Band Radio waves are defined by their frequency or wavelength. Different radio waves are then allocated different rules and uses. When a range of radio

---

5. SI stands for Système International (or International System in English), which is a system of standardized unit designations, typically in relation to scientific, engineering, and technical measurements such as seconds, meters, kilograms, and so on.
6. http://www.industrie.gouv.fr/metro/aquoisert/etymol.htm

frequencies are grouped together using the same rules, this group of frequencies is called a Radio Band.

Slave A simple device that works with a master. These devices are typically single-purpose devices.

Wi-Fi A complementary wireless technology that is designed for high data rates to connect computers and other very complex devices with the Internet.

Chapter 2
Basic Concepts

*In protocol design, perfection has been reached
not when there is nothing left to add,
but when there is nothing left to take away.*
—*IETF RFC 1925, Rule 12*

To understand Bluetooth low energy is to understand how low power consumption can be achieved in a short-range wireless system. The most basic design decisions are all built around enabling low power consumption in typical use cases.

Bluetooth low energy is not trying to optimize Bluetooth classic; instead, it is targeted at new market segments that haven't previously used open wireless standards. These market segments are those that require devices to send a few octets of data from once a second to once every few days. These are monitoring and control applications that perform tasks, such as detecting whether windows are open or closed for Smart Home heating applications, turning on and off appliances in response to electricity price fluctuations, or changing to a different TV channel.

2.1 Button-Cell Batteries

Button-cell batteries are the primary design goal for Bluetooth low energy. These batteries (see Figure 2–1) have very strict limits on how they can be used. The figure shows a CR2032, although other battery sizes are also available. The "CR" part of

Figure 2–1 A button-cell battery.

11

the battery label indicates it's a 3-volt battery made using lithium manganese dioxide. The "20" denotes that the battery is 20mm in diameter, and "32" specifies that it's 3.2mm in height.

For such small batteries, the maximum energy that can be stored in one manufacturer's battery is very similar to any other. A typical CR2032 will have a quoted energy capacity of 230mAh at 3 volts. Just to put this into context, this is about the amount of energy required to power a human being for just over 20 seconds. So by the time you've read this paragraph, you'll have used more energy than a typical CR2032 has available to power a Bluetooth low-energy device for a few years.

Even though 230mAh is a comparatively tiny amount of energy, a device will never be able to obtain all of it. First, the energy available is dependent on the temperature of the battery. The colder the battery, the less energy will be available. A button-cell battery at 0°C would only be able to provide 80 percent of the energy that is available at room temperature.

Second, if the battery is used aggressively, the total energy available will be significantly reduced. Typically, most button-cell batteries have a peak current that should not be exceeded without damaging the battery. This is typically 15mA. If this high-level current is drawn from the battery for extended periods of time, the total energy available would be reduced. Therefore, any successful radio design would need to manage this and allow the battery time to recover after a large or long current draw.

Finally, the battery itself has its internal leakage current that must also be taken into account. Even if a device is drawing no power from the battery, the battery is still losing charge. When the battery is used sparingly, the leakage current will start to become significant in the total energy budget.

2.2 Time Is Energy

Another basic concept that is used throughout the design of Bluetooth low energy is that time is energy. If the radio is doing something, even if it's nothing more than checking whether it needs to send or receive something, it's using energy. Consequently, it is important to reduce the time required to do anything useful.

A number of important and repetitive actions must be optimized. These include robustly discovering devices, connecting to devices, and sending data. By reducing the time required for these activities, the energy consumption is reduced, lengthening the life of the battery.

Robust device discovery requires a minimum of two devices: a device that is looking for other devices, and one or more devices that are discoverable. In Bluetooth low energy, for a device to be discoverable, it must transmit a very short message three times every few seconds, and if it needs to see if any other device wants to talk to it, it must listen immediately after broadcasting its message. A device that is looking for devices opens up its receiver and listens for devices that are transmitting.

Three transmits are done because three frequencies are used for robustness. The number three is chosen as a compromise between robustness and low power. If the number of frequencies was just one, like a lot of other technologies, then as soon as that frequency is blocked, the whole system would break. If the number of frequencies was, for instance, 16, the device would spend so much time just transmitting that it would not be low power anymore.

The choice about which device transmits and receives is also very deliberate. To search for a device that is transmitting requires you to listen for a long period of time; this uses a lot of energy and therefore should be done on the device with the bigger energy budget or a good reason to use the low-energy device. In Bluetooth low energy, the discoverable devices transmit, and the devices that are looking for the other devices receive.

The packet itself is very short. Short packets are good for three reasons. First, by using efficient encoding, short packets can send the same quantity of data faster, using less energy. Second, by restricting the devices to only use short packets, the requirements to constantly recalibrate the radio within the controller as the packet is transmitted are removed. Radios, when transmitting or receiving, will heat up, changing the characteristics of the silicon chip, and therefore changing the frequency of the transmissions. If the packets are kept short, the chip doesn't have enough time to heat up; thus, this energy-expensive procedure can be ignored. In addition, the requirement for a short packet also reduces the chip's peak power consumption a little. Finally, you can get more energy out of a button-cell battery by taking it in short-duration bursts and not a long continuous draw of current. Therefore, using several short packets with a sufficient space between them to allow the battery time to recover is better for the battery than using one longer packet.

2.3 Memory Is Expensive

Everybody knows that the more memory a computer has, the more expensive it is. However, every little bit of memory in a computer not only costs money but also costs energy. Memory typically requires dynamic refreshing—every so often the memory in the chip is refreshed. This dynamic refreshing requires energy. So the more memory that a device requires, the more energy is required to power the device. Therefore, the whole of Bluetooth low energy has been designed to reduce the amount of memory that is required in every layer.

For example, keeping the packets small in the Link Layer helps because it reduces the memory requirements for the radio when transmitting and receiving packets. For example, the Attribute Protocol Layer does not require any packets larger than 23 octets to be processed. It also does not require any state information to be saved between transactions. All this reduces the memory needed to do something useful.

Another burden for memory is the multitude of protocols that are required to be active when a device can do multiple things. For example, imagine a headset that does hands-free, remote control, and battery status reporting. If each of these use cases required a separate protocol, the memory required for each of those protocols would have to be added together. In Bluetooth low energy, there is only one protocol. The Attribute Protocol is used for name discovery, service discovery, and for reading and writing information required to implement a given use case. By having only one protocol, the overheads of multiple protocols are significantly reduced.

2.4 Asymmetric Design

One of the obvious design concepts in Bluetooth low energy, once the architecture has been understood, is the asymmetry that is evident at all layers. This asymmetric design is very important because the device with the smaller energy source is given less to do.

At the Physical Layer, there are two types of radios: transmitters and receivers. A device can have both a transmitter and a receiver. However, a device can implement only a transmitter or only a receiver. If one device only has a transmitter, and the other device only has a receiver, this is an asymmetric network.

This asymmetric design is all based on the fundamental assumption that the most resource-constrained device will be the one to which all others are optimized.

At the Link Layer, devices are divided into advertisers, scanners, slaves, and masters. An *advertiser* is a device that transmits packets; a *scanner* is a device that receives the advertiser's packets. A slave is connected with a master, but even here the asymmetry is evident. A slave cannot initiate any complex procedures, whereas a master has to manage the piconet timing, adaptive frequency hopping set, encryption, and a number of other complex procedures. The slave only does what it is told and doesn't have to perform complex processing at all. This keeps the slave very simple and therefore low cost, low memory, and using the lowest possible power.

At the Attribute Protocol Layer, the two types of devices are called client and server. The server holds data and the client sends requests to the server for this data.

The server, like the slave at the Link Layer, just does what it is told. The client has the hard job of working out what data the server has and how to use it.

Even the security architecture for low energy is asymmetric. The security architecture works on a key distribution scheme by which the slave device gives a key to the master device for it to remember. The burden is on the master to remember this bonding information; the slave doesn't have to remember anything. This means that it's simple for a slave device to support security, yet for a master device it is more complex.

This all implies that the most resource-constrained devices will want to be advertisers, slaves, and servers. These types of devices have the lowest possible memory and processing burden; therefore, the asymmetric design is beneficial to the goal of ultra-low power consumption on these types of devices.

The other types of devices—scanners, masters, and clients—have lots of resources to play with. These devices are typically associated with larger batteries, rich user interfaces, and possibly even an electric supply. It is right to move the burden then from the slave to the master, from the advertiser to the scanner, and from the server to the client. This reduces the power consumption of the most resource-constrained devices, to the cost of the most resource-abundant devices.

2.5 Design For Success

So many wireless standards fall down at the first hurdle because a great radio design just doesn't work when it starts to become popular due to congestion from many other radios. If there is one thing that Bluetooth does well, it is operating in a very congested environment. Three times a year, the Bluetooth Special Interest Group (SIG) organizes UnPlugFest testing events at which engineers from many competitive companies come together to test their devices before they are released to the market. These events highlight the fact that Bluetooth still works even when hundreds and even thousands of wireless devices come together in a single hotel ballroom. And Bluetooth low energy has learned from this.

Designing for success means that every person who gets on a crowded commuter train or bus or goes to a busy sports stadium or concert should be able to operate several low-energy devices. This means that thousands of devices could be within a few meters of a device, and device discovery and connections should function as expected. It also means that there should be no inherent limit to the number of devices a given device can talk to at the same time. If a device wants to talk to another device, then it should just be able to do that, not worry that there are only seven possible slaves that can be connected at the same time, which is the limit imposed by classic Bluetooth.

Device density is just one metric that was used during the design of the controller. Another was the security system. Any very popular radio system will become a target for people who want to try to break its security. This becomes even more important when monetary value is involved. So state-of-the-art security and encryption engines must be used.

Beyond security, if a person is going to be carrying around many devices, all of which are resource constrained and therefore advertising continuously, the issue of privacy must also be addressed. In Bluetooth low energy, privacy is dealt with as a major design goal. Every connection that is made uses a different signature that has no relation to any identifying information of these two devices. It is not possible to know who is walking down the street by just listening to the packets being transmitted during connections. Also, when advertising, it is possible to use a *private address*, which is a resolvable address that allows a friendly device to resolve the address if they have the identity resolving key but denies unfriendly devices the ability to resolve or track the address.

Another factor taken into account was that when the radio is used everywhere, even a single bit error can become significant. If you have a sewage outflow valve, for example, protecting your nice public park from being swamped by effluent, you really don't want to have a single bit error that causes that sewage outflow valve to open when you really wanted to make sure it was still closed. To protect against this, all packets have a strong cyclic redundancy check (CRC) value that can protect against all 1, 2, 3, 4, 5, and all odd bit errors. Also, if you want more robustness, you can start encryption. Then a different message authentication code is appended to the data to ensure that the data was sent from the device that you think it was sent from; no attacker will be able to reply to messages to the sewage outflow valve. And for the really vigilant, at the Attribute Protocol Layer, there is the ability to prepare a write into an attribute and only perform the execute of that write after the value to be written has been returned and verified. This means that for this single bit state for the valve, a total of 14 octets of CRC and authentication codes protect this data. Bluetooth low energy is robust.

2.6 Everything Has State

One of the basic concepts behind Bluetooth low energy is that everything has state. This state is exposed by using the Attribute Protocol in an attribute server. The state could be anything: the current temperature, the state of the battery of the device, the name of the device, or the description of where the temperature is being measured.

State doesn't just have to be readable state; it can also be written. A thermostat can have a set-point temperature by which another device can set the temperature to which this room should be heated or cooled. If you can expose state, you can also expose the state of a state machine. By using explicit state machine attributes, the state of the device can be clearly exposed. This offers the capability for clients to disconnect whenever they want, because when they reconnect, they can quickly determine the current state by just reading it.

Some state is variable and can change frequently. To enable an efficient transfer of state information from the server to the client, direct notifications of the state information from the server to the client are possible. These notifications don't require the client to poll the server, allowing very efficient application designs. The battery state could be notified only when something interesting happens; thus, a device wouldn't need to worry about the battery state at all until the notification arrives.

This simple-state–based model makes it possible for a very efficient client server architecture to be constructed. This also allows an object-oriented approach to state to be designed into applications, with reusable data types and service behavior. This reduces the quantity of code that a device needs to contain, which consequently reduces the power consumption of the device because memory doesn't need to be provided for that code. And there is another significant benefit to having less code: fewer bugs. Simpler systems are cheaper and faster to develop. A simpler system also typically contains fewer errors, making it more robust. Finally, simpler systems are easier to maintain. As Robert Browning said, "Less is more."

2.7 Client-Server Architecture

The client-server architecture has one additional basic design element that is fundamental to the design of Bluetooth low energy. When low energy was being designed, the problem of connecting devices to the Internet was considered. It could have been possible to put an Internet Protocol (IP) stack on every single resource-constrained device and just expose all the devices over the Internet. Unfortunately, even the simplest of IP stacks takes more memory and energy than is desirable on the simplest of devices. Therefore, the decision was made to not allow any IP packets to be routed directly to slave devices.

Instead, smart gateways allow the interconnection between the Internet and very efficient low-energy slaves. This interconnection is possible because of the pure client-server architecture. A server is just a repository of data, and it does not care who the client is. A client could be directly connected to the server or it could be connected via an Internet gateway from the other side of the planet.

This affords the ability for individuals to monitor and control their home when they are on vacation. And given that low energy will be used for everything from security alarms to set-top boxes and heating systems, it would be possible to check that all the windows are secure on the way to the beach, set up a recording of your favorite television program while you're lying back in the sand, and then turn the heat back up while flying home.

The ability to connect with gateways also allows sports and fitness devices to instantly update their associated web sites with their collected data, even before the exerciser has had a chance to finish her drink of water. It would also provide the ability to monitor elderly people so that they can stay in their own homes, safe in the knowledge that people are available should they need help.

The client-server gateway model also enables full Internet security to be used from the client to the gateway and allows the gateway to perform access control, firewall, and authorization of the client before granting it access to anything beyond the gateway. These are proven technologies used today in many homes and businesses.

2.8 Modular Architecture

One basic concept that is often overlooked is future-proofing the architecture. Most wireless standards are created in a rush, trying to get the technology out as quickly as possible, without much concern about how the technology will function in 10 or 20 years' time. This causes problems, because poor architectural decisions made under the duress of "time to market" damage the long-term viability of the platform. To solve this, the Bluetooth SIG created a special architecture working group just for the Generic Attribute Profile–based architecture to ensure a future-proof design.

The main outcome of this group has been the modular service architecture that builds on top of the generic attribute profile. This allows atomic encapsulatable bits of behavior to be wrapped up in a single service and exposed on a device in a standard way. (In this context, an atomic service is one that just does a single thing, and encapsulatable means that it can be separated from other functions and wrapped up by itself.) These services can reference other services, so a battery service can reference a temperature service if the battery has its own temperature sensor; this same temperature service can be reused for a home thermometer, a freezer temperature sensor, or a car engine coolant temperature sensor.

An interesting side effect of this architecture is that the services exposed on a device do not have to be directly related to a given profile. Profiles will require a given set of services on a given device, but that is about as much of a link as needed. This means that if another profile can be created to use a different combination of services on a device, it can combine the existing services in a different way without a problem. This is true, even if that profile was written after the services were designed and implemented in a device.

This is a highly flexible and modular architecture that can enable the building up of ecosystems over time. For example, smart meters could be deployed into homes to allow current and future price information and usage information to be exposed. Later, smart appliances can be deployed that allow themselves to be remotely turned on and off; using the gateway model, this could be controlled outside the home. Even later, a smart energy broker can be deployed that uses the information from the smart meter and the information from the smart appliances to save the homeowners money by scheduling energy use that takes into account the pricing information from the meter.

2.9 One Billion Is a Small Number

Any new technology faces a serious challenge trying to obtain market traction. For a technology to be successful, it has to be low cost. To be low cost, you need volume. To have volume, you need to be successful. Today, the single largest consumer electronics device that is sold is the cell phone. Any technology that makes it into the cell phone will be successful. Bluetooth is the classic example of this. Bluetooth low energy builds on Bluetooth's attach rate in cell phones to create an instant market.

The technology has the opportunity to have over one billion devices in the field within the first couple of years as the cell phone manufacturers update their platforms to include Bluetooth low energy. The interesting thing about this is that it creates a huge market for accessories for phones. And it is not just phones that can have Bluetooth low energy designed in quickly; computers, televisions, in fact, any devices that have Bluetooth classic, are likely to be updated to add Bluetooth low energy because of the extremely low cost associated with incorporating the new technology to an existing Bluetooth system.

2.10 Connectionless Model

Bluetooth classic was all about cable replacement: headset cables, mouse cables, file transfer cables. This implies an architecture where the cost of setting up a link is not that important because the link will be maintained for a few minutes, hours, or even days. The odd second delay at the start of the connection is not that important. Bluetooth low energy changes all this.

The basic concept with low energy is that connections are transient. When you need to do something or check something, you quickly create a connection, do what needs to be done, and then disconnect. A device that is only notifying some state information once every five minutes would only need the radio on for less than one second a day. This means that the radio is off 99.999 percent of the time; or to four significant digits, the radio is off 100 percent of the time. Any delay in each connection setup will cause a significant increase in power consumption.

Bluetooth low energy can create a connection, send data, and gracefully disconnect in about three milliseconds. This means that many devices that have some state information, but until now couldn't afford to add wireless technology because of the cost of energy requirements, can finally consider adding Bluetooth low energy. Even something as simple as a button can be enabled, possibly using scavenged power, and therefore never need a battery.

2.11 Paradigms

Most successful technology is built around sets of paradigms, and Bluetooth low energy is no different. Bluetooth low energy uses two main architectural paradigms: client-server architecture and service-oriented architecture.

2.11.1 Client-Server Architecture

The client-server architecture is a paradigm by which clients can send requests to servers over a network and the servers send back responses. It is the main paradigm behind the Internet, which is arguably the most successful networking technology ever released.

For example, when you type a URL address into a web browser, it first sends the address to a DNS server. This server responds with the IP address of the server that has been assigned to that name. The client then sends a hypertext transport protocol (HTTP) request to that server and, once connected, sends a request for the server to get the resource identified in the request. The server then responds with the appropriate resource, typically a text file that contains markup (HTML) information about how to display the information.

This file can also include additional URLs with which the client can fetch other resources, such as pictures or other pages. These additional links are really the reason HTML pages are thought of as being linked together into a *web*, hence the terms *web page* and *web server*.

There is a clear distinction between what a server does and what a client does. The server has information, typically in a structured form. This data is really why the server exists. This data can be anything, the current weather in Kona, Hawaii, the time of the next train from Seoul to the airport, or just some inane chatter between friends. The client, on the other hand, doesn't have any data. It just sends requests to servers. Once it receives the replies from a server, it can carry out the task it was assigned to do, such as display information to the user or notify the user that somebody they know has posted something on a wall or tweeted.

The main benefit of the client-server architecture is this defined split between the client and the server. This split is necessary when the different parts of the system are on different devices. By defining one of these parts as a server and one as a client, the explicit relationship between these two parts of the system can be determined.

The main benefit of this architecture is that it can scale. A client doesn't need to know anything except the URL to be able to access a resource. There can be many clients. Some sites on the Internet will have millions of requests made each day from millions of clients. The server doesn't really care what or where these clients are; it just responds to each request as it comes in.

This server architecture can also be scaled. A single machine responding to millions of requests a day might become overloaded and start to fail. The solution is to place many identical servers, all having access to the same information. This is further assisted when clients are given multiple IP addresses for a single name so that the load is spread evenly among each server. This is known as *load balancing*.

2.11.2 Service-Oriented Architecture

A further abstraction on top of the client-server architecture is the service-oriented paradigm. This is a model that organizes the information in a server into services. These services can be discovered, interacted with, and used with known semantics. This means that the services have a defined behavior that will always produce the same result, given the same preconditions.

This paradigm is the foundation of the most highly successful Internet systems, such as SOAP, REST, COBRA, RPC, Web Services, and so on.

A way to illustrate this is to relate it to a real-world example. Suppose that you have a package that needs to be delivered to another company quickly. The first thing you will probably do is call a courier company, arrange a pickup of the package, and then pay for the service. The key concept is that you know what is going to happen. There is an implicit set of behaviors that the courier company will be following. On any day, given the same package to be delivered, the courier company will do exactly the same thing—deliver it to its destination, on time. This service has known semantics with defined behavior that produces predictable results.

An interesting part of this is that you are interacting with two different people at the courier company: the person who answered the phone and took your request and the delivery driver who collected your package. You also used, although unknowingly, a third person who dealt with the financial transaction. Each of these people provide a subservice that, when combined, results in the primary service offering of the courier company.

Some of these subservices are also generic; they could be used interchangeably by many different types of companies. The processing of financial transactions is something that is done pretty much the same way in every company. Similarly, the function of taking phone calls for picking something up at one location and dropping it off at another could also be applied to taxi companies.

For all this to work, everything must adhere to a set of rules and conventions that are outlined in the sections that follow.

2.11.2.1 Formal Contract

For a service to be considered a service, it must follow a formal description of both its exposed functionality and how it behaves. For example, the courier company driver wears the company uniform, drives the company vehicle, and greets the customers in a pleasant manner. He will also drive between locations quickly and safely, and deliver the packages intact. Any violation of these rules would break the contract that the customer has with the courier company. Most courier companies therefore also require customers to agree to this formal contract before picking up their packages.

A side effect of having a formal contract is that it becomes easy for one instance of a service to be replaced by another instance of the service. This is possible only if the two instances of the service both expose the same functionality and behavior. For example, if the financial person left the company, it should be easy to find a replacement who knows the same accounting rules.

In Bluetooth low energy, these formal contracts are captured in service specifications that are formally adopted by the Bluetooth SIG. These specifications also have test specifications that ensure that the behavior of an implementation is valid.

2.11.2.2 Loose Coupling

In object-oriented software, each individual component of the system is meant to be designed as a separate object with no side effects. Those interactions that do occur between components can then be explicitly defined and tested.

By reducing the dependancies to a minimum, each service implementation can be changed without risk that unexpected side effects are either introduced or lost. Taking this to its logical conclusion, there should be a separation between the formal contract and its implementation. This then allows the implementation to be changed at will, as long as the formal contract is not broken and is unchanged.

For example, it is possible to add more drivers to the courier company, changing the implementation from a single driver doing everything to many drivers collecting packages from a small area of a city, bringing them back to a central warehouse, and then sending them back out, possibly with a different driver. From the customer's point of view, the service is identical, packages are collected and delivered as required, but the implementation is completely different. And this was possible without changing the financial or order-taking services.

2.11.2.3 Abstraction

Service abstraction is an import design rule because of the consequences if this rule is not followed. If there were no abstraction and a client had full knowledge of how a given service was implemented, that client might start to use that service in a way that constrains how the service can evolve.

It is common knowledge that more information is good. However, in the context of a service-oriented architecture, the less knowledge that a client has about how a service is provided, the better it is. With too much knowledge, the client might obstruct the reuse or redesign of the service because the client is implicitly linked to the specific implementation. If the service implementation changes, the client might break.

To ensure that this rule is followed, only the absolute minimum of state must be exposed by the service. Also, only the external manifestation of the service behavior should be specified.

2.11.2.4 Reusability

The concept of reusability is a design goal that was one of the promises of object-oriented methodologies for many years. However, reusability is really the ability for a service to be designed so that it could be applied to multiple different applications. Without careful thought, it is always easier to design a service that only does one job. With good design, services can be designed to be independent of the actual process that is used. This means that the service can then be reused in another application, quickly and easily.

The Bluetooth SIG has responded to this challenge by setting up a working group that has but one job, which is to look out for generic functionality and then abstract the requirements to enable significant reuse.

2.11.2.5 Statelessness

To be able to scale services for many clients, the servers cannot hold any client state data. A service could be defined that remembers everything the client has told them so that the client doesn't have to repeat this information on subsequent requests. The problem with this approach is that this information takes up a lot of memory and relies on this shared state information being in sync on both the client and server. This therefore leads to the server being reliant on the correct functioning of the client; this is a bad assumption.

Therefore, the statelessness design goal removes all the state from the interaction between the client and the server. There will still be some state information stored in the service, but this is always the server's state and is never the client's state. This means that any client can then send any request at any time and the server will respond to the same request in exactly the same way, regardless of which client made the request.

2.11.2.6 Composability

All the preceding goals imply that services should be designed to be both small and very simple. But the real world is never that simple. Real-world services are complex. To resolve this apparent conflict, a service-oriented architecture encourages the aggregation of smaller services to enable higher service interfaces.

The aforementioned design goals for services all encourage that they can be combined. For example, the courier company service was shown to be composed of three separate services. As long as each of these individual services followed the stated goals, these can be combined into a courier company. Similarly, some of these services could be combined to make a taxi company or an executive car service. The implementation might be different—a delivery van, a family car, or an executive limousine—but the services are essentially the same.

2.11.2.7 Autonomy

For services to be reused and combined, they must be reliable. A service that relies on other components within the system to perform received requests will not be as reliable as a service that has complete control over everything it does.

An autonomous service can stand alone and perform its task, regardless of what is going on around it. These services can be reused in other applications with very little difficulty. A service that is not autonomous would probably have to bring many other support services; these additional services might conflict with other services.

For example, the courier company drivers function autonomously, collecting and delivering packages, as they are told to do. They will continue to function autonomously even if there is a major malfunction in the office.

2.11.2.8 Discoverability

Finally, for services to be used, they must be discoverable. This might appear obvious at first glance, but service discoverability is essential for ad hoc networking. Without service discovery, all services must be statically programmed; a complex, burdensome, and error-prone task.

Typically, these use a separate protocol from the protocol used to interact with the service. For example, to find the courier company, somebody might have used a phone book or searched on the Internet; they would not just call random telephone numbers and hope that one would be a courier company.

Bluetooth low energy takes a different approach and uses a single protocol for both the discovery of services and interacting with these services. This protocol is called the Attribute Protocol, and service discoverability is described in its profile, the Generic Attribute Profile. Both of these are described in Chapter 10, Attributes.

Chapter 3
Architecture

> *That's been one of my mantras—focus and simplicity.*
> *Simple can be harder than complex: You have to work hard*
> *to get your thinking clean to make it simple. But it's worth it*
> *in the end because once you get there, you can move mountains.*
> —Steve Jobs

The architecture for Bluetooth low energy is fundamentally very simple. As shown in Figure 3–1, it is split into three basic parts: controller, host, and applications. The controller is typically a physical device that can transmit and receive radio signals and understand how these signals can be interpreted as packets with information within them. The host is typically a software stack that manages how two or more devices communicate with one another and how several different services can be provided at the same time over the radios. The applications use the software stack, and therefore the controller, to enable a use case.

Within the controller, there is both the Physical Layer and Link Layer as well as a Direct Test Mode and the lower layer of the Host Controller Interface. Within the host are three protocols: Logical Link Control and Adaptation Protocol, Attribute Protocol, and the Security Manager Protocol. Also within the host are the Generic Attribute Profile, the Generic Access Profile, and modes.

3.1 Controller

The controller is the bit that most people can identify as the Bluetooth chip or radio. Calling the controller a radio, however, is very simplistic; the controller is composed of both analog and digital parts of the radio frequency components as well as hardware to support the transmission and reception of packets. The controller interfaces with the outside world through an antenna and to the host through the Host Controller Interface.

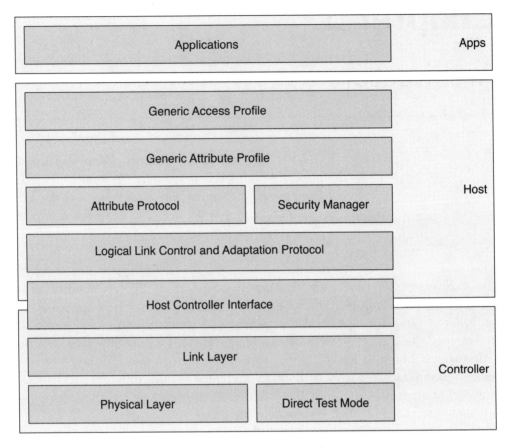

Figure 3–1 The Bluetooth Architecture

3.1.1 Physical Layer

The Physical Layer is the bit that does the hard work of transmitting and receiving bits using the 2.4GHz radio. To lots of people, the Physical Layer is magical. Fundamentally, it is not magic, but is the simple transmission and reception of electromagnetic radiation. Typically, radio waves can carry information by varying the amplitude, frequency, or phase of the wave within a given frequency band. In Bluetooth low energy, the frequency of the radio waves are varied to allow either a zero or a one to be exposed, using a modulation scheme called Gaussian Frequency Shift Keying (GFSK).

The Frequency Shift Keying part means that ones and zeros are coded onto the radio by slightly shifting the frequency up and down. If the frequency is shifted abruptly to one side or the other at the moment the frequency changes, there is a pulse of energy that spreads out over a wider range of frequencies. So a filter is used to stop the energy spreading too far into higher or lower frequencies. In the case of GFSK, the filter used is shaped like a Gaussian curve. The filter used for Bluetooth low energy is not as tight as the filter used for Bluetooth classic. This means that the low energy radio signal spreads out a little more than the classic radio signal.

This slight widening of the radio signal is useful because it means the radio comes under spread-spectrum radio regulations, whereas the Bluetooth classic radio is governed by frequency-hopping radio regulations. Spread-spectrum radio regulations allow a radio to transmit on fewer frequencies than frequency-hopping radio regulations. Without the more relaxed filter shape, the Bluetooth low energy radio would not be allowed to advertise on just three channels; it would have to use many more channels, which would make the system higher power (as discussed earlier).

This slight widening of the radio signal is referred to as the modulation index. Modulation index describes how wide the upper and lower frequencies used are around the center frequency of a channel. When the radio signal is transmitted, a positive frequency deviation of more than 185kHz from the center frequency represents a bit with the value 1; a negative frequency deviation of more than 185kHz represents a bit with the value 0.

For the Physical Layer to work, especially when lots of other radios are in the same area transmitting at the same time, the 2.4GHz band is split up into 40 separate RF channels, each 2MHz apart from one another. The Physical Layer transmits information at one bit of application data every one microsecond. For example, to send the 80 bits of data for the string "low energy" formatted in UTF-8 would take just 80μs, although this does not take into account any packet overhead.

3.1.2 Direct Test Mode

Direct Test Mode is a novel approach to the testing of the Physical Layer. In most wireless standards, there is no standard way to get a device to perform standard Physical Layer tests. This leads to the problem of many different companies building their own proprietary methods to test only their Physical Layers. This increases the costs for the whole industry and increases the barriers for an end-product manufacturer to change from one silicon supplier to another quickly.

Direct Test Mode allows a tester to command a controller's Physical Layer to either transmit a sequence of test packets or receive a sequence of test packets. The tester can then analyze the packets received, or the number of packets that the device under test received, to determine if the Physical Layer is working according to the specification. The tester can also measure various RF parameters from received packets to determine if the Physical Layer is compliant with the RF specs. The Direct Test Mode is not just applicable to qualification testing; it can also be used for production line testing and calibration of radios. For example, by quickly commanding a Physical Layer to transmit on a given radio frequency, and measuring the actual transmitted signal, the radio can be tuned to match what it should be doing. This sort of calibration is typically done on every single unit, so having test equipment that can do this efficiently can save product manufacturers money.

3.1.3 Link Layer

The Link Layer is probably the single most complex part of the Bluetooth low energy architecture. It is responsible for advertising, scanning, and creating and maintaining connections. It is also responsible for ensuring that packets are structured in just the right way, with the correctly calculated check values and encryption sequences. To do this, three basic concepts are defined: channels, packets, and procedures.

There are two types of Link Layer channels: advertising channels and data channels. Advertising channels are used by devices that are not in a connection sending data. There are three advertising channels—again, this is a compromise between low power and robustness. Devices use these channels to broadcast data, advertise that they are connectable and discoverable, and to scan and initiate connections. The data channels are only used once a connection has been established and data needs to flow. There are 37 data channels, and they are used through an adaptive frequency-hopping engine to ensure robustness. The data channels allow data from one device to another to be sent, acknowledged, and, if necessary, retransmitted. Data channels can be encrypted and authenticated on a per-packet basis.

To send data on any of these channels (data or advertising), small packets are defined. A packet encapsulates a small amount of data that is sent from a transmitter to a receiver over a very short period of time. Packets include information to identify the intended receiver, as well as a checksum that ensures that the packet is valid. The basic packet structure is identical between the advertising channels and data channels, with a minimum of 80 bits of addressing, header, and check information included in each and every packet. Figure 3–2 presents an overview of the Link Layer packet structure.

The packets are optimized to increase their robustness by using an 8-bit preamble that is sufficiently large to allow the receiver to synchronize bit timing and set

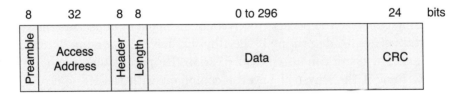

Figure 3–2 The Link Layer packet structure

the radio's automatic gain control; a 32-bit access address that is fixed for advertising packets but completely random and private for data packets; an 8-bit header to describe the contents of the packet; an 8-bit length field to describe the payload length, although not all these bits are used for length because no packet with more than 37 octets of payload is allowed to be sent; a variable-length payload that contains useful data from the application or the host device stack; and finally, a 24-bit cyclic redundancy check (CRC) value to ensure that there are no bit errors in the received packet.

The shortest packet that can be sent is an empty data packet that is 80μs in length, whereas the longest packet is a fully loaded advertising packet that is 376μs in length. Most advertising packets are just 128μs in length, and most data packets are 144μs in length.

3.1.4 The Host/Controller Interface

For many devices, a Host/Controller Interface (HCI) will be provided that allows a host to communicate with the controller though a standardized interface. This architectural split has proven to be extremely popular in Bluetooth classic, for which over 60 percent of all Bluetooth controllers are used through the HCI interface. It allows a host to send commands and data to the controller and the controller to send events and data to the host. It is really composed of two separate parts: the logical interface and the physical interface.

The logical interface defines the commands and events and their associated behavior. The logical interface can be delivered over any of the physical transports, or it can be delivered via a local application programming interface (API) on the controller, allowing an embedded host stack to be included within the controller.

The physical interface defines how the commands, events, and data are transported over different connection technologies. The physical interfaces that are defined include USB,[1] SDIO,[2] and two variants of the UART.[3] For most controllers, they will support just one or possibly two interfaces. It should also be considered that to implement a USB interface requires lots of hardware, and the interface is not the lowest power interface, so it would not typically be provided on a Bluetooth low energy single-mode controller.

Because the host controller interface has to exist on both the controller and the host, the part that is in the controller is typically called the lower-host controller interface; the part that is in the host is typically called the upper-host controller interface.

---

1. USB stands for Universal Serial Bus.
2. SDIO stands for Secure Digital Input Output.
3. UART stands for Universal Asynchronous Receiver/Transmitter.

3.2 The Host

The host is the unsung hero of the Bluetooth world. The host contains multiplexing layers, protocols, and procedures for doing lots of useful and interesting things. The host is built on top of the upper-host controller interface. On top of this is the Logical Link Control and Adaptation Protocol, a multiplexing layer. On top of this are two fundamental building blocks for the system; the Security Manager that does everything from authentication and setting up secure connections and the Attribute Protocol that exposes the state data on a device. Built on the Attribute Protocol is the Generic Attribute Profile that defines how the Attribute Protocol is used to enable reusable services that expose the standard characteristics of a device. Finally, the Generic Access Profile defines how devices find and connect with one another in an interoperable manner.

There is no defined upper interface for the host. Each operating system or environment will have a different way of exposing the host APIs, whether that be through a functional or object-oriented interface.

3.2.1 Logical Link Control and Adaptation Protocol

The Logical Link Control and Adaptation Protocol (also referred to as L2CAP) is the multiplexing layer for Bluetooth low energy. This layer defines two basic concepts: the L2CAP channel and the L2CAP signaling commands. An L2CAP channel is a single bidirectional data channel that is terminated at a particular protocol or profile on the peer device. Each channel is independent and can have its own flow control and other configuration information associated with it. Bluetooth classic uses most of the features of L2CAP, including dynamic channel identifiers, protocol service multiplexers, enhanced retransmission, and streaming modes. Bluetooth low energy just takes the absolute minimum of L2CAP.

In Bluetooth low energy, only fixed channels are used: one for the signaling channel, one for the Security Manager, and one for the Attribute Protocol. There is only one frame format, the B-frame; this has a two-octet length field and a two-octet channel identifier field, as illustrated in Figure 3–3. This is the same frame format that classic L2CAP uses for every channel until the frame formats are negotiated to something more complex. For example, in Bluetooth classic, it is possible to have frame formats that include additional frame sequencing and checks. These are not needed in Bluetooth low energy because the checks at the Link Layer are strong enough to not need additional checks, and the simple Attribute Protocol has no need for out-of-order delivering of packets from multiple channels. By keeping the protocols simple and doing sufficient checks, only one frame format was required.

3.2 The Host

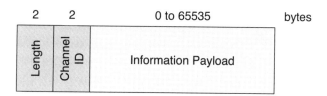

Figure 3-3 The L2CAP packet structure

3.2.2 The Security Manager Protocol

The Security Manager defines a simple protocol for pairing and key distribution. Pairing is the process of attempting to trust another device, typically by authenticating the other device. Pairing is typically followed by the link being encrypted and the key distribution. Using key distribution, shared secrets can be distributed from a slave to a master so that when these two devices reconnect at a later date, they can quickly prove their authenticity by encrypting using the previously distributed shared secrets. The Security Manager also provides a security toolbox for generating hashes of data, generating confirmation values, and generating short-term keys used during pairing.

3.2.3 The Attribute Protocol

The Attribute Protocol defines a set of rules for accessing data on a peer device. The data is stored on an attribute server in "attributes" that an attribute client can read and write. The client sends requests to the server and the server responds with response messages. The client can use these requests to find all the attributes on a server and then read and write these attributes. The Attribute Protocol defines six types of messages: 1) requests sent from the client to the server; 2) responses sent from the server to the client in reply to a request; 3) commands sent from the client to the server that have no response; 4) notifications sent from the server to the client that have no confirmation; 5) indications sent from the server to the client; and 6) confirmations sent from the client to the server in reply to an indication. So, both client and server can initiate communication with messages that require a response, or with messages that do not require a response.

Attributes are addressed, labeled bits of data. Each attribute has a unique handle that identifies that attribute, a type that identifies the data stored in the attribute, and a value. For example, an attribute with the type Temperature that has the value 20.5°C could be contained within an attribute with the handle 0x01CE. The Attribute Protocol doesn't define any attribute types, although it does define that some attributes can be grouped, and their group semantics can be discovered via the Attribute Protocol.

The Attribute Protocol also defines that some attributes have permissions: permissions to allow a client to read or write an attribute's value, or only allow access to the value of the attribute if the client has authenticated itself or has been authorized by the server. It is not possible to discover explicitly an attribute's permissions; that can only be done implicitly by sending a request and receiving an error in response, stating why the request cannot be completed.

The Attribute Protocol itself is mostly stateless. Each individual transaction—for example, a single read request and read response—does not cause state to be saved on the server. This means that the protocol itself requires very little memory. There is one exception to this: the prepare and execute write requests. These store a set of values that are to be written in the server and then executed all in sequence, in a single transaction.

3.2.4 The Generic Attribute Profile

The Generic Attribute Profile sits above the Attribute Protocol. It defines the types of attributes and how they are used. It introduces a number of concepts, including "characteristics," "services," "include" relationships between services, and characteristic "descriptors." It also defines a number of procedures that can be used to discover the services, characteristics, and relationships between services, as well as read and write characteristic values.

A service is an immutable encapsulation of some atomic behavior of a device. This is a long stream of very complex words, but it is a very simple concept to understand. Immutable means that once a service is published, it cannot change. This is necessary because for a service to be reused it can never be changed. As soon as a service's behavior changes, version numbers and other awkward setup procedures and configuration take time and therefore become the antithesis of a connectionless model, one of the basic concepts behind Bluetooth low energy.

Encapsulation means expressing features of something succinctly. Everything about a given service is enclosed and expressed through a set of attributes in an attribute server. Once you know the bounds of a service on an attribute server, you know what information that service is encapsulating. Atomic means of or forming a single irreducible unit or component of a larger system. Atomic services are important because the smaller the server, the more likely it is to be reusable in another context. If we created complex services that had multiple, possibly related behaviors, the chance of these being reused is significantly reduced.

Behavior means the way something acts in response to a particular situation or stimulus. For services, the behavior means what happens when you read or write an attribute, or what causes the attribute to be notified to the client. Explicitly defined behavior is very important for interoperability. If a service is specified with

3.2 The Host

poorly defined behavior, each client might act in a different way when interacting with the service. The services might then act differently depending on which client is connecting, or more important, the same service on different devices will act differently. As soon as this becomes entrenched in the devices, interoperability is destroyed. Therefore, explicitly defined behavior that is testable, even for erroneous interactions, promotes interoperability.

Service relationships are key to the complex behaviors that devices expose. A service is atomic by nature. Complex behaviors should not be exposed in just a single service. Take, for example, a device that can measure the room temperature by exposing a temperature service. The device might be powered by a battery so it would expose a battery service. However, if the battery also has a temperature sensor, we should be able to expose another instance of the temperature service on the device. This second temperature service needs to be related to the battery so that a client can determine that relationship. This is shown in Figure 3–4.

To accomodate complex behaviors and relationships between services, services come in two types: primary services and secondary services. The type of a service is not typically dependent on the service itself but on how that service is used in a device. A primary service is one that exposes what the device does, from the perspective of the user. A secondary service is one that is used by a primary service or another secondary service to enable it to provide its complete behavior. In the previous example, the first temperature service would be a primary service, the

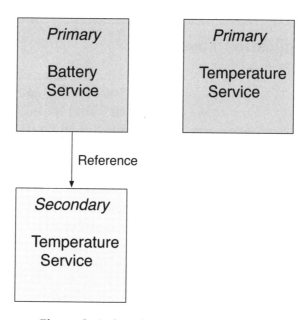

Figure 3–4 Complex service relationships

battery service would also be a primary service, whereas the second instance of the temperature service—the temperature of the battery—would be a secondary service referenced from the battery service.

3.2.5 The Generic Access Profile

The Generic Access Profile defines how devices discover, connect, and present useful information to the users. It also defines how devices can create a permanent relationship, called *bonding*. To enable this, the profile defines how devices can be discoverable, connectable, and bondable. It also describes how devices can use procedures to discover other devices, connect to other devices, read their device name, and bond with them.

This layer also introduces the concept of privacy by using resolvable private addresses. Privacy is important for devices that are constantly advertising their presence so that other devices can discover and connect to them. Devices that want to be private, however, must broadcast by using a constantly changing random address so that other devices cannot determine which device it is by listening, or which device is moving around by tracking its current random address over time. However, to allow devices that are trusted to determine if it is nearby, and to allow connections, the private address must be resolvable. The Generic Access Profile, therefore, defines not only how private addresses are resolvable but also how to connect to devices that are private.

3.3 The Application Layer

Above the controller and the host is the Application Layer. The Application Layer defines three types of specifications: characteristic, service, and profile. Each of these specifications is built on top of the Generic Attribute Profile. The Generic Attribute Profile defines grouping attributes for characteristics and services, and the applications define the specifications that use these attribute groups.

3.3.1 Characteristics

A characteristic is a bit of data that has a known format labeled with a Universally Unique Identifier[4] (UUID). Characteristics are designed to be reusable, and therefore have no behavior. As soon as behavior is added to something, it limits its reuse. The most interesting thing about characteristic specifications is that they are defined in

---

4. UUIDs are documented as part of ITU-T Rec. X.667 | ISO/IEC 9834-8, which is duplicated in the IETF as RFC 4122.

a computer-readable format rather than as human-readable text. This gives computers the ability, when they see a characteristic used for the first time, to download this computer-readable specification and use it to display these characteristics to the user.

3.3.2 Services

A service is a human-readable specification of a set of characteristics and their associated behavior. The service only defines the behavior of these characteristics on a server; the service does not define the client behavior. For many services, the client behavior can be implicitly determined by the service's server behavior. However, for some services, there might need to be more complex behavior in the client that must be defined. This client behavior is defined in profiles, not in the services.

Services can include other services. The parent service can only define the services that are included; it cannot change the characteristics in these included services or change the behavior of these services. The including service, however, can describe how multiple included services interact with each other.

Services come in two variants: primary and secondary, as noted in Section 3.2.4. The primary or secondary nature of a service can be defined in a service specification or can be left up to the profile or an implementation. Primary services are those that embody what a given device does—it is these services that the user would understand that the device does. Secondary services are those that assist the primary services or other secondary services.

Services do not describe how devices connect to each other to find and use services. Services only describe what happens when a characteristic is read or written, or when it is notified or indicated. Services do not describe what Generic Attribute Profile procedures are used to find the service, the characteristics within a service, or how the characteristics are used by a client.

3.3.3 Profiles

Profiles are the ultimate embodiment of a use case or application. Profiles are specifications that describe two or more devices, with one or more services on each device. Profiles also describe how the devices should be discoverable and connectable, thereby defining what topology is necessary on each device. Profiles also describe the client behavior for finding the service, finding the characteristics of the service, and using the service to enable the functionality required by the use case or application.

There is a many-to-many mapping of profiles to services, as illustrated in Figure 3–5. A service can be used by many profiles to enable a given behavior on a device. The behavior of a service is independent of which profile is using this service at this time. Application stores can be given the list of services that a device

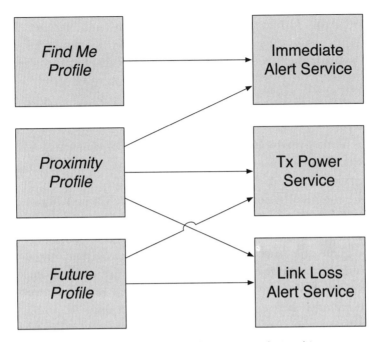

Figure 3–5 Complex profile service relationships

supports and find the set of applications from the store that use these services. This flexibility enables a plug-and-play model that has worked so well for the universal serial bus.

3.4 Stack Splits

It is possible to build a Bluetooth low energy product by using multiple different stack splits. The specification defines one stack split using the host controller interface between the controller and the host, but you can use many other different stack splits.

3.4.1 Single-Chip Solutions

The simplest stack split that is possible with Bluetooth low energy is the single-chip solution, as shown in Figure 3–6. This has no stack splits; all parts of the product are packed into a single chip. This chip includes the controller, the host software, and the applications. This is the ultimate in low-cost products, only requiring a source of power, an antenna, some hardware to interface to—for instance, buttons and lights—and some additional discrete components.

Unfortunately, there are some downsides to using single-chip solutions. First, the development environments are more difficult to use because the chips are very

3.4 Stack Splits

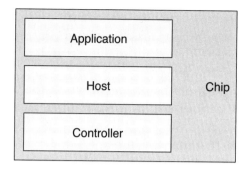

Figure 3–6 A single-chip solution

resource constrained. Second, to reduce the cost, the software needs to be burned into read-only memory (ROM) in the chip. This requires a custom chip to be made for a single product. This can be offset by the reduced bill of materials for very large production runs, but the process can be very expensive for smaller production runs.

For devices that have small production runs or for prototype products, a mass-produced single chip that includes everything from the controller up to the top of the host can be used with a small nonvolatile memory chip to store the application. This yields very low-cost small production runs. At power up, the contents of the nonvolatile memory are read into the single chip and executed. Thus, you can have both an efficient prototyping platform and a cost-effective small production run product.

3.4.2 Two-Chip Solutions

For two-chip solutions, the classic model is that the controller is on one chip and the host and applications are on a separate chip, as shown in Figure 3–7A. This model is typically used for cell phones and computers because they already have very

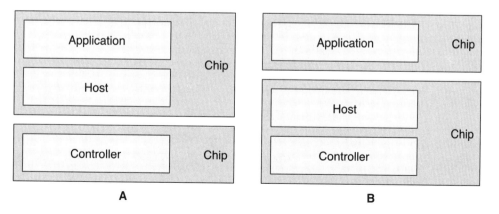

Figure 3–7 A pair of two-chip solutions

powerful processors capable of running the complete host and application software stack. This solution typically uses mass-produced controller chips with a standard Host Controller Interface. Although this architectural split is ideal for devices that already have a very powerful processor, it is not ideal for any other type of device.

An alternative two-chip solution is one in which the controller and host are on one chip, and the applications are on a separate chip, as shown in Figure 3–7B. This has the advantage that the application chip can be a very small low-power microprocessor because the application chip doesn't need much memory or other resources to run the application. The interface between the two chips would typically be a custom interface, probably employing a simple UART. This solution has the advantage that two standard mass-produced chips can be combined and use the standard development tools for the application chip.

3.4.3 Three-Chip Solutions

It is also possible for multiple-chip solutions to be used. For example, it would be possible to combine a standard controller on one chip, with a host chip and an application chip, as shown in Figure 3–8. The host chip would require two separate interfaces.

These solutions are typically prohibitively expensive; therefore, they are typically confined to development systems where multiple interfaces are used to allow each layer to be separately instrumented. Short production runs might also be able to tolerate this complexity because the cost of integrating into fewer components is not offset by the savings on each end product if only a few products are ever to be manufactured. For mass-produced products, this architecture would never be viable from a cost point of view.

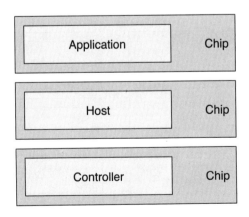

Figure 3–8 A three-chip solution

Chapter 4
New Usage Models

All of the books in the world contain no more information than is broadcast as video in a single large American city in a single year. Not all bits have equal value.
—*Carl Sagan*

Bluetooth low energy enables a new way of using wireless technology. The main new models are based around the advertising model and include presence detection, broadcasting of data, and connectionless models. They also include gateways from devices to the Internet.

4.1 Presence Detection

The most interesting new wireless model enabled by Bluetooth low energy is presence. Presence means the state or fact of existing, occurring, or being present in a place or thing. Using the advertising model, devices can passively scan in the background for other devices that are broadcasting. The devices that are advertising might be just advertising their address, or they could be advertising some presence-based data.

Advertising is a new mode of operation defined in the Link Layer. With it, devices can periodically transmit their identity and a small amount of information. This model is possible because the modulation index of the radio has been increased and the 2.4GHz band regulations permit wider radio signals to be sent using a non-frequency-hopping radio. Because the radio doesn't employ frequency hopping, fewer channels are needed for devices to be connectable and discoverable. This means that it is much more efficient to advertise and scan.

Scanning is possible in two modes: active and passive. Active scanning requires the scanner to request more information from advertisers, to obtain additional static data. Passive scanning just requires the scanner to listen for advertising packets. Once an advertising packet is received by the Link Layer, it can be sent to the host.

The host can use the information about what devices are nearby to determine where it is. For example, if the host discovers a car, then the host can determine that it is in or near a car and change its behavior accordingly, perhaps by connecting to the car. Similar use cases around the home, in the office, or at a café are also possible. It is this automatic determination in the background using passive scanning that allows a device to automatically change its behavior based on where it is.

Presence, as just described, is about a mobile device determining where it is. Another type of presence is about static devices being able to determine what devices are in a given location. Probably the most useful benefit of this would be to find somebody or something in a large office building, for example. The devices or people that want to be tracked would advertise infrequently, and devices in each room would monitor which devices they can detect. This information can then be communicated to a central device to determine location. You can use this to automatically route phone calls to the nearest phone or to track employees during an emergency evacuation of a building.

4.2 Broadcasting Data

The advertising model also allows a small quantity of data to be broadcast—a very small amount, just a few tens of octets of data—but the ability to broadcast this small bit of information to any device that is listening in the area is incredibly valuable. As stated earlier, the ability to determine where a device is, based on what devices are broadcasting, is a useful function in its own right. However, this relies on having some way to map the device that is broadcasting to a physical location. Data broadcasting helps with this mapping.

You can use broadcasting to transmit many different types of useful data. There are three main areas for which broadcasting data can help the user experience: initial connection setup, advertising, and broadcasting information.

To help with an initial connection setup, devices can broadcast data about what type of device they are and that they want to connect to a device with a complementary set of services or profiles. For example, when you remove a television from its box and switch it on for the first time, it starts to search for a remote control. When you install the batteries in the remote control, it starts to advertise that it is looking for the television.

The television receives this broadcast data, connects to the remote control, automatically pairs with it, and then allows the remote control to talk to it securely. This means that, from the consumer's perspective, they turn on the television for the first time, put the batteries in the remote control, and then press buttons on the remote to control the television; no connect buttons and no pairing menus.

Advertising is a useful tool for many organizations. With it, consumers have the ability to discover real world-services from over 100 meters away. An obvious place where advertising using a free wireless technology is useful is at international airports and railway stations. The ability to advertise gate or track details for flights or trains gives travellers who don't want to spend lots of money on roaming charges or Wi-Fi Internet access an alternative way to gather information. Simple bus stops can also advertise when the next bus will arrive and where the bus is heading.

It is also possible to broadcast information that is gathered locally by a device. For example, a temperature sensor could broadcast the temperature to any device that is currently listening for temperature information. This is most useful when information is being sensed that changes rapidly and the information is useful for multiple devices.

4.3 Connectionless Model

One of the biggest changes from Bluetooth classic to Bluetooth low energy is in the way that a connectionless model has been designed and implemented. In a connectionless model, devices do not need to maintain a connection for useful information to be exchanged quickly between them. Because the main protocols never establish a connection-oriented channel between devices, there is no cost to dropping and then reconnecting a connection when data needs to be sent. This encourages devices to only establish a connection when they need to send data, and not to maintain an expensive connection just in case some data does need to be sent. This connectionless model does impose some interesting design changes from standard wireless protocols.

In a connection-oriented channel, the state information can be established over a period of time by using the protocol. The state information, therefore, typically is not available whenever it is required, but only by remembering the state that has been implicitly created by both devices. This state information requires a long time to be established, causing delays upon the initial connection while the state information is discovered and negotiated. Protocols that are based on implicit state typically have negotiation and configuration procedures as well as feature bits and version numbers. If a connection is going to be up for a long time, and there is a lot of state information, that state-full system can be more efficient.

Unfortunately, many protocols are not fully defined, with each bit of state implicitly defined as opposed to being explicitly defined. This leads to interoperability problems because each device thinks that the connection has a different state and therefore makes different assumptions about what can or should happen next. This is one of the biggest problems with connection-oriented systems. This can be solved by defining the state explicitly and also how any state machines work. A good example of this would be the Logical Link Control and Adaptation Protocol (L2CAP) Layer in Bluetooth where for Bluetooth classic, a simple state machine and configuration system are used when establishing a connection. All the state of the connection is explicitly defined, and the connection state machine is fully described. This, however, has taken over 10 years to develop to the exemplary level it now occupies.

Thus, the connectionless model solves these problems by not defining the state of a connection, but the state of the device. By exposing state through a stateless protocol, such as Attribute Protocol, it is possible to disconnect at any time and, upon reconnection, determine what the current state is directly from the other device. It is also possible to explicitly define state machines with both an exposed state and an exposed control point to persuade that state machine into different states as defined by some service. It is also possible to reestablish a connection just because some information in this state has changed and a device has registered to receive this state change information.

For example, it can be used to signal the battery level of a device. A monitoring device would connect to the battery-powered device, read the current battery level, configure the battery level to be notified when it changes, and then disconnect. When the battery level does change, the battery-powered device slowly makes itself connectable and the battery-monitoring device notices that it has something to say. The battery-powered device then connects to the battery-monitoring device. Immediately after establishing a connection, the battery-powered device can notify the monitor of the new battery level and then immediately terminate the link. This can all happen within about 3 milliseconds. For a device that was fully charged and has a battery that lasts for just one year, this would require approximately 99 reports of 3 milliseconds each, a total radio-active time of just fewer than 300 milliseconds. In contrast, just to set up a connection-oriented channel in Bluetooth classic can take a similar time, *for each report.*

4.4 Gateways

The most radical change in computing technology over the last few years has been the spread and pervasiveness of the Internet. It appears that everything is connected to it, from newspapers to televisions and radios. All of these are portals for media, whether it is printed words, moving pictures, or voice. Of course, there are many other uses for the Internet, including communications such as e-mail and social networking, and teleconferencing audio and video links. However, the next big challenge will be to connect hundreds more devices for each device that's connected to the Internet today. This is a big change and the current infrastructure will probably not have the capacity to cope with all this new data initially.

4.4 Gateways

The problem with the Internet is that it is built around a connection-oriented model. A TCP[1] connection is a session-oriented channel that is established between two devices, which takes time to set up. The fact that devices connected to the Internet have to have an address is also session based. To obtain an address, a device must either have this address programmed into it or it has to ask another device to allocate it an address for a period of time; it can use this address for this period of time, after which it must ask for another address.

The biggest problem with the Internet as it is currently structured is that it is designed around a wired infrastructure. Wires are great; they are mostly reliable, and because the devices connected with wires aren't moving around, they can also be connected to other wires supplying electricity. This means that energy efficiency for wired protocols is rarely considered. The fact that routers constantly check for the mapping of allocated or nonallocated Internet addresses to devices, at stochastic intervals, typically means that Internet devices need to be listening all the time. This doesn't work when devices are constantly moving around and need ultra-low power consumption. Another approach has to be used.

The model followed in Bluetooth low energy is one that is used in most homes that have more than one computer connected to the Internet. This is the concept of a gateway using network address translation (NAT). To the outside, your typical home has a single Internet address, allocated to the gateway or router. The gateway, however, allocates a separate set of addresses for all the devices in the home that are attached to it. The key is that the gateway translates the internal addresses to the single external address, hiding the topology of the internal network from the outside world. The outside world just sees one device, and it doesn't really care about which device is really sending or receiving the data.

The Internet Protocol itself is very expensive. For an IPv6-based network, a 128-bit source and destination address has to be included in every single packet that is transmitted. This means that the minimum size of a packet, before any other protocol overhead, is 32 bytes. This is larger than the biggest Bluetooth low energy packet. Therefore, it becomes very difficult to just use the Internet Protocols directly over low energy—even if we were to discount the fact that they were designed when everything was wired. The gateway model, however, allows us to hide the internal addressing of devices from the outside world. This internal addressing of devices could be using a separate IPv6 address space or could be using some other addressing scheme that is transparent to the outside world.

---

1. TCP stands for Transmission Control Protocol. This is the protocol that is used to transmit a stream of data between two devices.

By using Bluetooth low energy gateways, tiny wireless devices not only can connect to the Internet, but they can do so using the least possible amount of power. Bluetooth low energy does this by pushing the complexity of the Internet to the gateway devices that have the resources to cope. It also allows these gateway devices to map Internet addresses to devices using any scheme they want. This could be done by allocating an individual IPv6 address to each device or by using a port number of a single Internet address to each device.

Gateways are useful because if your refrigerator needs to notify its manufacturer that the compressor pump is failing and needs to be replaced under warranty, it must have a way to get this message out to the manufacturer's Internet server. Obviously, similar things will need to be done for all manufactured devices: washing machines, cars, and vacuum cleaners, to name just a few. The manufacturer might also want to send information to these devices; for example, they might want to upgrade washing programs. Thus, gateways provide the way for devices to interact with the Internet without being burdened by the power-hungry wired protocols that drive the Internet.

Part II

Controller

Chapter 5, The Physical Layer, describes how devices communicate wirelessly with one another.

Chapter 6, Direct Test Mode, highlights the Direct Test Mode and its role in performing low-cost, time-efficient testing of the radio.

Chapter 7, The Link Layer, introduces the lowest layers of protocol that describe packets, advertising, and how to create a connection.

Chapter 8, The Host/Controller Interface, describes the interface that hosts can use to talk with a controller and get it to do useful things.

Chapter 5
The Physical Layer

You see, wire telegraph is a kind of a very, very long cat. You pull his tail in New York and his head is meowing in Los Angeles. Do you understand this? And radio operates exactly the same way: you send signals here, they receive them there. The only difference is that there is no cat.
—*Albert Einstein*

5.1 Background

Two or more Bluetooth low energy devices use radio waves to send and receive information among them. Radio has been around for many years, starting with very simple spark-gap transmitters, evolving through amplitude modulation and frequency modulation, and recently, to phase-shift keying and other more complex modulation schemes.

The sections that follow provide an introduction to how radios work, from the basics up to modern modulation schemes that are used by Bluetooth low energy.

5.2 Analog Modulation

The basic spark-gap radio is possibly the simplest radio that you can build. It is so simple that you could build it with just two components. Take a nine-volt battery with the two battery terminals at the top of the battery and a metal coin that can conduct electricity. But before you make your transmitter, you need to set up a receiver; you can use a radio that is turned to the AM band but not to any particular radio station. Then, briefly contact the coin to the battery terminals. You should hear the radio pick up the interference caused by the coin and the battery as the electricity sparks across the gap when the coin is very close to the terminals at the top of the battery but not actually touching them.

There are two problems with spark-gap radios. First, they are not very efficient. You need a very large electrical potential difference to transmit a long distance, typically many thousands of volts. Second, there is only one radio able to transmit at the same time in the same area. This severely restricts the ability to communicate more than one message at the same time in any given region. For example, imagine the residents of an entire city having the "choice" of just one television station.

The next advance in radio technology was amplitude modulation, or AM radio. It was observed that you could transmit a single frequency by using radio waves. These carrier signals could then be modulated to transmit some information. In amplitude modulation, the amplitude, or volume, of the carrier signal was changed. Fundamentally, this was a huge advance because many different radio signals could be transmitted at the same time. Figure 5–1 shows a representation of an analog amplitude modulation signal.

Countries and private companies would go on to create many different radio stations. When short-wave radio was the only type of radio available, governments would set up their radio station and just pick an empty part of the spectrum, or deliberately pick a part that would interfere with another country's station so that their citizens could listen to the local propaganda only. Eventually, international agreements were created to allocate frequencies in a logical and nonconflicting manner. It is these agreements that have been the basis of most radio frequency allocations since.

Sometimes, the allocation of frequency bands might be very similar, but how they are used would be different. For example, the medium-wave radio bands used for AM radio in the United States and the European Union are both between 530kHz and 1620kHz, yet in the United States, each station is allocated a frequency at 10kHz intervals, whereas in the European Union they are at 9kHz intervals. This means that the European Union can have more stations, but radio receivers must be designed to cope with both different frequency bands.

Amplitude modulation also has problems that are self evident when you listen to an AM radio station. If the audio input to a radio station is very quiet, the receiver might either lose the signal completely or it will output more noise as it desperately attempts to receive something useful. This noise always exists; it is called *background noise*, and it is generated by the many electrical devices that exist in our world. It is also caused by lightning and other atmospheric effects, including radiation from the sun.

The next advance in radio transmission would significantly increase the sound quality by removing the effect of the background noise from the signal. This was done by using frequency modulation, or FM radio. Instead of modulating the carrier with amplitude, so that when the input is very weak the output carrier is very weak,

Figure 5–1 Analog amplitude modulation

Figure 5-2 Analog frequency modulation

the frequency of the carrier is instead modulated with the input (see Figure 5-2). Whereas audio is the input signal, this means that a very quiet input would cause virtually no deviation of the carrier frequency, but a very loud input would cause a large deviation in the carrier. The most important thing about frequency modulation is that the carrier is always transmitted at maximum power so that a receiver can lock onto the signal and then demodulate the information out of the signal. The other main advantage of frequency modulation is that many more carriers can be placed in very close proximity to each other. A modern FM transmission, for example, would have a mono signal (Left + Right), a stereo signal (Left/Right), digital information about this station (Radio Data System), as well as pilot tones.

Frequency modulation therefore solves both the main problems with spark-gap radios and amplitude modulation. It is relatively simple, has the ability for a receiver to lock on to the signal regardless of what that signal is, and has the ability to have much closer grouped signals. There are yet more advanced modulation schemes, such as phase modulation. Phase modulation is similar to frequency modulation in that the frequency of the signal changes based on the input. Phase modulation makes changes to the phase of the signal. Phase modulation is more complex than simple frequency modulation and is typically used in complex digital systems only, such as digital radio. Beyond this, there is quadrature amplitude modulation, which uses amplitude modulation on two different carriers that are 90 degrees out of phase with one another. Quadrature amplitude is used for transmitting digital television in most of the world.

5.3 Digital Modulation

Before discussing digital modulation, it is useful to quickly recap the difference among chips, bits, and symbols. The input data is expressed as bits, which have a value of either zero or one. One bit can be combined with other bits to form a multiple-bit value. These combined bits are collectively called a symbol. A symbol is therefore one value that can represent multiple bits. There is the other way, although rarely used in real life, when a bit is actually transmitted by using multiple chip codes. Each chip is actually a fractional bit; thus a single bit is made up of many chips. On the excessively complex end of the scale, it is even possible to combine multiple chips into a single symbol that represents multiple bits.

When radio systems are compared, various numbers are normally bandied about. Most are compared with each other when they are not actually directly comparable. The most useful number is the *application data rate*. This is the maximum data rate at which application data can be transmitted after taking account of any packet overhead and the maximum rate at which packets can be transmitted. Packet overhead is any extra symbols that are needed to define and manage the transmission itself in addition to the actual application data. This can include timing synchronization information, addressing information, headers that describe the application data, and checks to ensure that the application data is valid when received.

The most often quoted number is the *physical bit rate*. This is the maximum number of bits that can be transmitted in one second if the radio were to transmit data continuously for that complete second. Apart from television and radio stations, very few radios can continuously transmit data. Instead they must split the data into multiple, self-contained packets. Another very useful number is the symbol rate. The *symbol rate* is the maximum number of symbols that can be transmitted in a second; this determines the speed at which the receiver must work. The higher the symbol rate, the more energy is required to process these symbols to extract the information bits.

Another piece of information that is important to understand to quantify how much application can be sent is the *frame rate*. This is the number of packets that can be sent within a given period of time. The radio must be turned around from being a transmitter to being a receiver. Turning the radio around takes time. The longer this turn-around time the less time is spent transmitting application data.

When transmitting digital information, the modulation schemes become much simpler to understand. The most simple digital modulation is the on-off keying, or OOK. Keying is a jargon word that describes how the carrier is adjusted for a given digital signal. On-off keying means that we take a carrier and either have it on or off at various times. This could be considered a very simple amplitude modulated signal with the input being full on or full off. Thus, as shown in Figure 5–3, if the absence of a carrier encoded the value zero, and a carrier encoded the value one, it is easy to see that it is possible to transmit 8 bits of information by just transmitting the appropriate amplitude of the carrier at the appropriate time. Although this is very simple, it is prone to noise, especially when the signal is very weak.

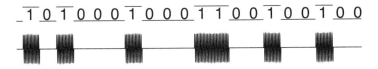

Figure 5–3 On-off keying

5.3 Digital Modulation

Next in complexity is *amplitude-shift keying*, or ASK (see Figure 5–4). It is analogous with amplitude modulation. If the input signal is binary—0 percent and 100 percent—ASK will degenerate to OOK. If it were possible to represent four different levels—0 percent, 33 percent, 66 percent, and 100 percent—this can encode two bits of information for each level. The 8 bits of data could then be transmitted twice as quickly as an OOK encoded system, in just 4 symbols. Both ASK and OOK have problems with low signal-to-noise ratios, such that making the determination as to whether the symbol represents anything other than zero becomes problematic.

Frequency-shift keying, or FSK, is the next step up in complexity. This is analogous with frequency modulation. As Figure 5–5 demonstrates, shift in frequency is used as the key to determine the symbol's value. The simplest frequency-shift keying is *binary frequency-shift keying*, for which an input bit of zero yields a negative frequency deviation, and a input bit of one yields a positive frequency deviation of the carrier. Using FSK, more levels can be used by varying rates of frequency deviation. The advantage of frequency deviation is that the carrier can always be received and therefore locked upon, allowing much longer range than the simpler ASK approach. Equation 5-1 shows a mathematical representation of frequency deviation.

$$h = \frac{\Delta f}{f_m} \quad (5\text{-}1)$$

The size of the deviation is called the modulation index h, where Δf is the maximum deviation from the carrier frequency, and f_m is the highest frequency in the source signal being modulated onto the carrier. If the modulation index is greater than one, the carrier is varying in frequency more than the source signal, then the signal is called a wide-band transmission. If the modulation index is less than one, the carrier is varying in frequency less than the source, then the signal is called a narrow-band transmission. A modulation index of 0.5 is considered a very special

Figure 5–4 Amplitude shift keying

Figure 5–5 Frequency shift keying

value because this is the value used for *minimum-shift keying*, or MSK. MSK is a variant of FSK that is very spectrally efficient.

5.4 Frequency Band

Bluetooth low energy uses the 2.4GHz Industrial, Scientific, and Medical (ISM) band for transmitting information. This frequency band is very special for two reasons. Most radio spectrum is licensed, meaning you have to buy a license to transmit anything. For example, your local radio station might have lots of money to buy a license to transmit music and advertising, on the premise that the revenue from the sales of commercial spots is sufficient to pay for the music, wages, and license. Some frequency bands are licensed but at zero cost. This includes bands for aviation, military, and civilian emergency services. Some bands are free. Yes, free!

The 2.4GHz ISM band is one of the license-free frequency bands. You do not need to buy a license to transmit in this band as long as you follow the rules. The rules are very simple; the band must be used for a personal area network (PAN) or a local area network (LAN) with limited range and limited transmit power. The details of the rules themselves are very complicated; however, in essence, it is free to use for short-range applications.

The second reason the 2.4GHz ISM band is very special is that it is the only license-free spectrum that is the same in every country. This means that no matter where you buy a product that uses this band, you can use it in any other country without having to configure it. There are other ISM bands, such as those around 900MHz, but these use different frequencies and band sizes, depending on the locations; the United States uses 915MHz, whereas the European Union uses 868MHz. The 2.4GHz band, which is useable anywhere, extends from 2400MHz to 2483.5MHz. This gives a total available spectrum of about 83.5MHz.

5.5 Modulation

The Bluetooth low energy radio uses *Gaussian frequency-shift keying*. A Gaussian filter is one that optimizes the transition from one symbol to the next by increasing the time that is used to slide the frequency from one value to another. Without this filter, the frequency shift would be dramatically quick, causing much noise to be created. This means that when changing from a zero bit to a one bit, the transition is both fast and efficient.

When transmitting data, Bluetooth low energy transmits at one million bits per second (Mbps), with one bit per symbol. The modulation index is approximately 0.5, meaning that it is very close to the optimal MSK. The modulation index can

5.6 Radio Channels

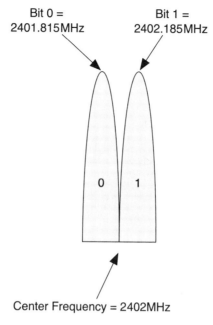

Figure 5–6 Modulation

vary between 0.45 and 0.55, meaning that Bluetooth low energy is not classified as an MSK radio; however, it has most of the MSK properties, including reducing side-band power output, which means less expensive filters are required to make it conform to the regulatory requirements.

To send a zero, a negative frequency deviation is used. To send a one, a positive frequency deviation is used. The minimum frequency deviation is about 180kHz. This means that if a center frequency of 2402MHz is used, a zero would be indicated by a transmission at 2401.820MHz, and a one would be indicated by a transmission at 2402.180MHz, as shown in Figure 5–6.

5.6 Radio Channels

If you want to implement the most robust, longest-range system, it is always best to use more than one frequency to transmit information. Some systems achieve this by having very wide transmissions; 802.11, for example, uses either 20MHz- or 40MHz-wide channels to send data very quickly. Other systems do this by frequency hopping; Bluetooth classic uses 79 narrow channels that are all used when transmitting information. The choice is mostly arbitrary, except that many narrow channels will be able to find a way through complex multi-path environments that are constantly changing much more efficiently than fewer wide-band transmissions.

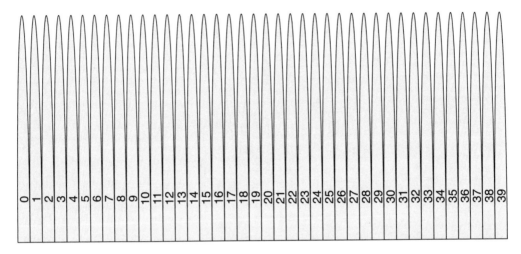

Figure 5-7 Radio channels

Figure 5-7 illustrates that Bluetooth low energy uses 40 radio channels to transmit information. The center frequency for each channel can be calculated very simply, as shown in Equation 5-2.

$$f_c = 2402 + 2k \tag{5-2}$$

Here, f_c is the center frequency of radio channel k.

This means that the lowest frequency used in Bluetooth low energy is 2402, and the highest frequency is 2480. At the bottom end of the frequency band, a gap of 2MHz is provided between a Bluetooth low energy channel and anything using the next frequency band below it. At the top end of the frequency band, a gap of 3.5MHz is provided between a Bluetooth low energy channel and anything using the next frequency band above it.

5.7 Transmit Power

In the 2.4GHz ISM band, there are limits to the maximum transmit power that a device can use to stay within the license-free regulations. For Bluetooth low energy, the specification limits the maximum transmit power to +10dBm. The LE specification also imposes that there is a minimum transmit power of −20dBm, so devices cannot be made so quiet that no other devices can hear them.

A + 10dBm transmit power means that it would be transmitted at 10mW, whereas at −20dBm, it would be transmitting at just 10µW.

5.8 Tolerance

All devices are manufactured with a given tolerance. Typically, the more accurate the tolerance, the more costly the devices. For the radio, the major tolerance that can be specified is the frequency accuracy. Even if a radio is designated to transmit around 2402MHz, it might actually be operating at 2401.850MHz or 2402.150MHz. This is the tolerance in the center frequency when transmitting a packet. In Bluetooth low energy, the center frequency tolerance is ±150kHz for the whole packet. The reason that the center frequency might be off is that it is typically obtained by multiplying the frequency from a known frequency crystal. This crystal would typically have a frequency of 16MHz; therefore, it must be multiplied by a factor of over 150 to get to up 2400MHz. Any inaccuracies in the crystal would be multiplied, as well, and included in the transmission frequencies. For example, if the crystal was actually outputting 16.0001MHz, the center frequency would be off by approximately 150kHz. This crystal would be said to have an error rate of 62 parts per million (ppm). Typically, low-cost, high-volume crystals with an error rate of approximately 50 ppm are readily available.

Another value that is very important is how much the radio drifts from its center frequency during the packet. This drift is caused by heat that builds up in a silicon chip during use. As the heat builds, the internal frequencies used in the radio will drift slightly. A Bluetooth low energy radio cannot drift more than 50kHz during a packet. This means that if the radio started transmitting perfectly at 2402.000MHz at the start of a packet, it would have to be between 2401.950MHz and 2402.050MHz at the end of the packet. There is also a maximum drift rate of 400Hz/µs.

5.9 Receiver Sensitivity

When building a receiver, there is really only one question that matters: How good is it? This is quantified by measuring the receiver sensitivity: how sensitive the radio is to detecting wireless transmissions from another device. This is measured in dBm, and is typically a very small number. The required receiver sensitivity for Bluetooth low energy is −70dBm. In other words, it has to be able to pick up 0.0000001mW of electromagnetic energy to be able to work. However, noise will always be present. There is no point in being able to detect a signal if you can't decode it. Therefore, in practice, the sensitivity threshold is set at the value where a signal can be decoded with an acceptable bit error rate (BER). For Bluetooth low energy, this has been chosen as 0.1 percent BER.

Most controllers supporting Bluetooth low energy will have a receiver sensitivity of about –90dBm, or 1pW. This is an incredibly small amount of energy that is able to be detected from the noise of the band, but this leads to impressive ranges, as explained in the following section.

5.10 Range

To calculate the range of a Bluetooth low energy radio, the *link budget* of the system needs to be determined. The link budget is made up of a number of elements that use the power from the transmitter in a silicon chip before it is received by a peer silicon chip. These elements include the antenna and matching circuit gains and losses. However, assuming that the antenna and matching circuits make little difference,[1] the main contributor to the link budget is the *path loss*. Path loss is a measure of how much the radio signal has reduced in power between the antenna in the transmitter and the antenna in the receiver. Equation 5-3 determines the *path loss* required for a given distance. Table 5–1 presents the correlation between path loss and distance; Figures 5–8 and 5–9 show the relationship graphically. It should be noted that this equation is an approximation, valid only for an isotropic antenna, and ignores any losses in the transmit/receive systems.

$$path\ loss = 40 + 25 log(d) \qquad (5\text{-}3)$$

In the equation, d is the distance between the transmitter and the receiver.

Table 5–1 The Relationship of Path Loss to Distance

Path Loss (*path loss*)	Distance (d)
50dB	2.5 m
60dB	6.3 m
70dB	16 m
80dB	40 m
90dB	100 m
100dB	250 m
110dB	630 m

---

1. Sometimes, this is an invalid assumption if the antenna and matching circuits are poorly designed. Good module manufactures or RF engineers should be able to reduce this problem.

5.10 Range

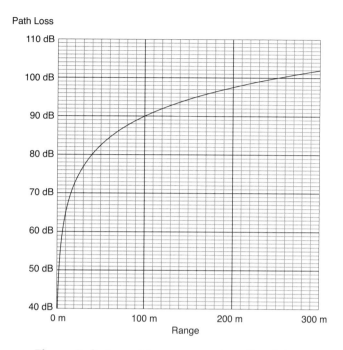

Figure 5–8 A graphic representation of path loss

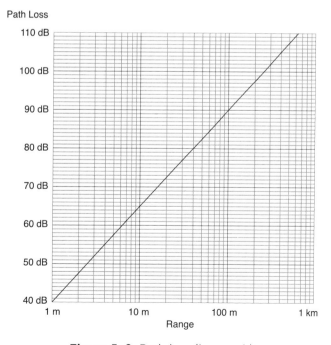

Figure 5–9 Path loss (log graph)

When the transmit power is −20dBm and the receiver sensitivity is −70dBm, a path loss of 50dB, the range is 2.5 meters. This is the distance possible when the minimum transmit power is used, with the minimum receiver sensitivity.

When the transmit power is 0dBm and the receiver sensitivity is −80dBm, a path loss of 80dB, the range is 40 meters. This is the distance possible when a moderate transmit power is used, with a moderate receiver sensitivity.

When the transmit power is 10dBm and the receiver sensitivity is −90dBm, a path loss of 100dB, the range is 250 meters. This is the distance possible when the maximum transmit power is used with the receiver sensitivity possible with modern chips.

Chapter 6
Direct Test Mode

Knowledge must come through action; you can have no test which is not fanciful, save by trial.
—Sophocles

6.1 Background

One of the biggest problems with wireless systems, especially those that are designed for the lowest possible cost of production, is how to calibrate them and perform qualification and product line tests of their performance. This is especially true after the device has been packaged into another module or product, and there is no way to move aside the host stack to perform a few seconds of testing at the start of the device's life. Direct Test Mode solves all these problems by defining standard testing procedures and a hardware interface to drive this protocol even after a host stack and other parts of the device have been incorporated in the device.

For the direct test mode to work, three devices are required (see also Figure 6–1):

- A Device Under Test (DUT)
- An Upper Tester (UT)
- A Lower Tester (LT)

The DUT is the controller, module, or end product that is being tested. The device must have both an antenna and a Universal Asynchronous Receiver Transmitter (UART) or Host/Controller Interface (HCI) to the UT.

The UT is typically manufactured by a test-equipment manufacturer and includes software to drive the device under test through the UART or HCI interface as well as the ability to communicate and drive the LT.

The LT is a device that can transmit and receive packets, effectively communicating with the device under test through the device's antenna.

The device under test is told what to do by the UT and transmits or receives packets. The UT at the same time informs the LT to do the opposite; that is, to

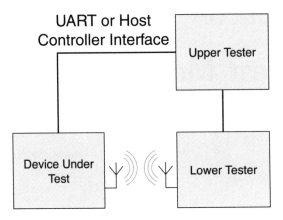

Figure 6–1 Test configuration

receive or transmit packets, respectively. This means that the device under test will transmit packets to or receive packets from the LT. At the end of the test, the UT can use the information available from both devices, a packet count from the device under test or more comprehensive information from the lower tester, to determine if the DUT passed the tests.

The UT can also do calibration of the controller on a production line by asking the device to transmit packets at a known frequency and measuring the actual frequency that the controller is transmitting. Typically, this is required if the external crystals used for a timing reference are not exactly at their design frequency. This crystal trimming would be done while the controller is transmitting packets, allowing very fast calibration of parts.

6.2 Transceiver Testing

To test the transceiver, the controller can either be asked to transmit or receive packets. When the controller transmits packets, it does so for a length of time determined by the tester. The tester then attempts to receive these packets and can determine various Physical Layer properties from these received packets. For example, the LT can measure the frequency drift of the device under test while transmitting. When the transceiver receives packets, it does so for another length of time determined by the tester. The tester sends a known number of packets, and the DUT just counts the ones that were correctly received. This information can be passed back to the UT when the test has completed. By doing so, the tester can determine how well the receiver has performed for a given number of transmitted packets.

6.2 Transceiver Testing

6.2.1 Test Packet Format

The packet format for test packets is very similar to the advertising packet format, as described in Chapter 7, The Link Layer, Section 7.3. The access address used is the bit inverse of the advertising access address. However, given that this access address is only ever used during testing, and can never be used in a product during a typical connection, it is not a protected value such as the main advertising access address.

The test packet format uses the advertising packet header format, so the first four bits are the test packet type, and all other bits are set to zero. The test packet types that are defined include:

- PRBS9 (a 9-bit Pseudo-Random Bit Sequence)
- "11110000"
- "10101010"
- PRBS15 (a 15-bit Pseudo-Random Bit Sequence)
- "00001111"
- "01010101"

Only the first three packet types are used in the qualification test specification. However, these tests might be useful for production line testing. For all advertising packets, whitening is disabled. This must be done to allow accurate measurement of frequency deviation for the nonrandom bit sequences.

6.2.2 Transmitter Tests

The transmitter tests determine how accurately the transmitter in the DUT is performing. You can use the transmitter tests to determine frequency deviation, frequency drift, and other radio parameters that are specified.

To start the transmitter test, the UT sends a command to the DUT. This command includes three parameters that determine the test to be performed. The first parameter is the frequency that will be tested. This is the radio channel number, as defined in Equation 5-2. The next parameter is the packet payload length; valid values are 0 to 37 bytes, inclusive. The last parameter is the type of data that will be transmitted.

Three types of data in test packets can be transmitted:

1. **PRBS9**—Used for power transmission testing
2. **11110000**—Used for frequency deviation testing
3. **10101010**—Used for carrier and initial frequency testing

The PRBS9 packet sequence is a pseudo-random bit sequence that uses a repeating 9-bit sequence. It is generated by using a linear feedback shift register. The PRBS9 sequence is commonly used as a test pattern to test the performance of the radio as quickly as possible. The primary reason for using the PRBS9 sequence is that it closely resembles the random nature of whitened packets used in connections. As such, it is easily used for power transmission tests as well as receiver sensitivity tests.

The 11110000 packet sequence is a repeating sequence of four ones and four zeros. This is used to test the frequency deviation when the same bit is transmitted continuously and then moved to the other bit. The tester will be looking for a frequency deviation of over 225kHz. This illustrates why the whitening must be turned off in test mode; otherwise, the radio would not transmit repeating bits.

The 10101010 packet sequence is a repeating sequence of one and zero. This is used to test the frequency deviation when alternate bits are transmitted. This is used to perform carrier testing and for measuring the initial frequencies used in transmissions.

After the LT has received enough packets to obtain a suitable result, the UT commands the DUT to stop the test. The DUT immediately stops transmitting and returns an event to confirm that it has finished transmitting. The UT can then start another test, possibly using a different packet type, a different packet length, a different frequency, or a combination of all three.

6.2.3 Receiver Tests

The receiver test is much simpler than the transmitter test. The receiver tests are used to determine the bit error rate at various transmit power levels. This requires the LT to transmit packets at a known transmit power, typically by using a conducted antenna connection to the DUT so that uncertainty of the path loss over the air is removed from the equation. The DUT counts the number of successfully received packets and sends this information to the UT at the end of the test. The UT can then determine the resulting number of packet errors, thereby estimating the bit error rate of the receiver at the given signal strength.

To start the receiver test, the UT sends a command to the DUT. The command includes just one parameter. The parameter is the frequency that will be tested, the radio channel number, as in the transmitter tests. When the DUT receives this command, the DUT resets a packet counter to zero and starts receiving.

When the receiver receives a valid packet, the packet counter is incremented. The DUT continues to increment until the UT commands the DUT to stop the test. Once the stop command is received, the DUT immediately stops receiving and returns an event to confirm that it has finished receiving. The event includes the number of valid packets received by the DUT. The UT can then compute the performance of the receiver, based upon how many packets were transmitted and how many were received. The UT can then start another test, possibly using a different frequency or at a different transmit power.

6.3 Hardware Interface

To enable very efficient and device-independent testing of modules, possibly from multiple controller manufacturers, a standardized hardware interface is defined that can be used by UTs. The hardware interface is a simple two-wire UART that has a single line for transmitting bits from the UT to the DUT, and another line for transmitting bits from the DUT to the UT. It is expected that module manufacturers and end-product manufacturers will expose two pins or pads that can be connected to the UT on a production line to allow for calibration and verification of an individual product's ability to transmit and receive.

6.3.1 UART

UARTs are typically very flexible in how they can be configured. To reduce the problem of an incompatible interface, the Direct Test Mode UART only has one configurable parameter: baud rate.

The baud rate for the DUT can be one of the following values: 1200, 2400, 9600, 14400, 19200, 38400, 57600, or 115200. The typical rate is 38400 because this is a good compromise between efficient command and event transfer and implementation cost.

The rest of the UART parameters are very standard. Each byte is 8 bits in length, sent with no parity, and one stop bit. There is no flow control, either software or hardware. Obviously, it is impossible to do hardware flow control because there are no hardware flow control wires. Software flow control is also not required because a command or event is only a maximum of 2 bytes in length and only one command can be sent at a time. The final design parameter is a common ground. It is assumed that there is a common ground between the two devices.

To send a command or event, 2 bytes of data must be sent a maximum of 5 milliseconds apart. Once a command is sent, an event must be returned within 50 milliseconds. This ensures that the DUT starts and stops tests in a timely manner, and therefore, the accuracy of any counting is easily validated.

6.3.2 Commands and Events

Four commands can be sent from the UT to the DUT:

- Reset
- Transmitter Test

- Receiver Test
- Test End

Two events can be sent from the DUT to the UT:

- Test Status
- Packet Reporting

The reset command does exactly as it says. It stops the controller at whatever point it is located; if the controller was transmitting or receiving test packets, it must stop. It also places the controller into a known good state. It immediately returns a test status event to confirm that everything is back to the quiescent state. Only 2 bits within the reset command are used; all other bits are ignored.

The transmitter test command starts the transmitter test. This includes the three parameters: frequency, length, and packet type, as shown in Figure 6–2. Once this command is received, the DUT sends a test status event and starts transmitting packets using the parameters. If for some reason the device cannot transmit packets, the DUT still returns the test status event but sets the status bit to denote an error. If the transmitter test command failed to start, the UT would need to reset the DUT to place it into a known good state.

The receiver test command starts the receiver test. This includes the frequency parameter; all other bits in this packet are ignored. Once this command is received, the test status event is returned and the DUT starts receiving packets on the desired frequency. Again, if the device cannot receive any packets, the test status event is returned with the status bit showing an error.

After a successful transmitter or receiver test, the test end command can be sent by the UT to stop the test. When this command is received by the DUT, it immediately stops what it was doing and returns the packet reporting event. If the receiver test was being run, this packet-reporting event includes the packet count for the successfully received packets (see Figure 6–3). If the transmitter test was being run, this field in the event is set to zero because no packets are received in the transmitter test. After the packet-reporting event is received, the UT can start another test by using another test command.

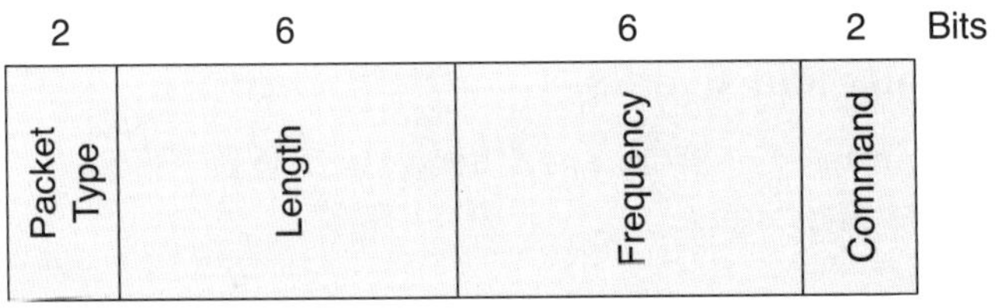

Figure 6–2 The Direct Test Mode command bit structure

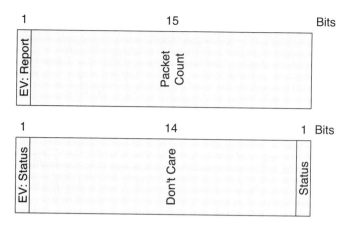

Figure 6–3 Direct Test Mode events reporting bit structure

The test status event only defines a single bit. If this bit is zero, the test command was successful. If this bit is one, the test command failed for some reason. There are two possible reasons a test command fails:

- The controller is in a state that confuses it, and a reset is probably the most suitable next step for the UT.

- The controller doesn't support the test, probably because it doesn't have either a transmitter or receiver. (Although it possible to test the receiver of a transmitter-only device.)

It is not possible to determine from this command set if the DUT supports a given test; a UT can only try each test and determine the status.

The packet-reporting event is used to report that the test completed. It includes a single 15-bit packet count of the number of successfully received packets. This packet count will always be zero for transmitter tests, but can be any value from 0 to 32,767 for receiver tests. The DUT doesn't worry about overflow of the packet counter; if the packet counter overflows, a test that might have taken a very long time can determine that only one packet was received, even though 32,769 packets were actually received. UTs should therefore only run tests for a duration that would not cause overflow of the packet counter.

6.4 Direct Testing by Using HCI

It is also possible to reuse the existing HCI transports (see Chapter 8, The Host Controller Interface, Section 8.2) and logical interface to exercise direct test mode on a controller. This does require more infrastructure to be in place, especially if the host interface is complex. However, for controllers that are being individually tested—as opposed to being within a highly optimized module design or end product—this is a very valid approach.

The test procedures are identical, except that instead of sending 2-octet commands and events, full HCI commands and events are sent. The mapping of the HCI commands and events to the Direct Test Mode commands and events is shown in Table 6–1.

There is no dedicated test status event or packet reporting event when using HCI because the Command Complete already performs this function, includes an opcode to determine which command triggered this Command Complete, and has differing parameters, depending on this command.

Table 6–1 HCI to Direct Test Mode

Direct Test Mode Command or Event	HCI Command or Event
Reset	HCI Reset
Transmitter Test	HCI LE Transmitter Test
Receiver Test	HCI LE Receiver Test
Test End	HCI LE Test End
Test Status	HCI Command Complete
Packet Reporting	HCI Command Complete

Chapter 7
The Link Layer

> *All this technology for connection and what we really only know more about is how anonymous we are in the grand scheme of things.*
> —*Heather Donahue*

The Link Layer defines how two devices can use a radio to transmit information between one another. This includes defining the detail of a packet, advertising, and data channels. It also defines procedures for discovering other devices, broadcasting data, making connections, managing connections, and ultimately sending data within connections. This is compounded by the challenges of wireless communication systems in the 2.4GHz ISM band, including interference, noise, and deep fades.

7.1 The Link Layer State Machine

Before we discuss packets and how they are used, it is important to understand the basic concept of the Link Layer state machine and its implications on the design of Bluetooth low energy.

As shown in Figure 7–1, the Link Layer state machine defines just five states:

- Standby
- Advertising
- Scanning
- Initiating
- Connection

However, it should be considered that the scanning state has two substates: active scanning and passive scanning. The connection state also has two substates: master and slave.

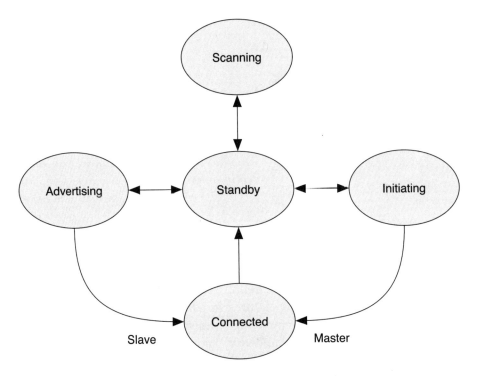

Figure 7–1 The Link Layer state machine

Although the Link Layer state machine explains how devices can discover and connect to one another, it also explains another fundamental design decision that Bluetooth low energy implemented; the separation of the broadcast, discovery, and connection processes from the data transmitted in a connection. Part of this design was done for ultra-low power consumption on the part of the advertising devices. By reducing the number of advertising channels to just three, you can maintain robustness while reducing power consumption. But this requires separate advertising states and separate advertising packets. The Link Layer state machine has three states in which advertising packets are sent or received, and one state in which data packets are sent and received.

7.1.1 The Standby State

When Link Layers are powered on, they start in the standby state and remain there until the host layers tell them to do otherwise. It is possible to move from the standby state into either the advertising, scanning, or initiating states (see Figure 7–2). It is also possible to move into the standby state from every other state. The standby state is really the center—the most important, albeit inactive, state.

7.1 The Link Layer State Machine

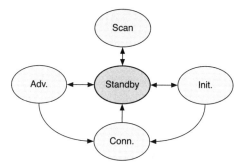

Figure 7-2 The standby state

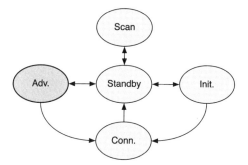

Figure 7-3 The advertising state

7.1.2 The Advertising State

The advertising state (see Figure 7-3) allows the Link Layer to transmit advertising packets. It can also respond to scan requests from devices that are actively scanning by sending a scan response. The advertising state is required if a device wants to be discoverable or connectable. The advertising state is also required if a device wants to broadcast data to other devices in the area.

To be an advertiser, a device *must* have a transmitter, but it might have a receiver as well. A device that only supports the advertising state could be built with just a transmitter, saving the cost of the receiver on that chip. It should be noted that in practice the volume pricing for a dual-purpose chip with both a transmitter and a receiver might well end up being less expensive than a low-volume transmit-only chip.

It is possible to move from the advertising state into the standby state by stopping advertising. It is also possible to move from the advertising state into the connection state when an initiating device sends a connect request packet to this advertiser.

7.1.3 The Scanning State

In the scanning state (see Figure 7–4), a device will receive advertising channel packets. This could be used to simply listen to see what devices are advertising in the local area. Scanning is composed of two different substates: passive scanning and active scanning. Passive scanning only receives advertising packets. Active scanning also sends scan requests to advertising devices to obtain additional scan response data.

It is only possible to move from the scanning state into the standby state. This is done by stopping scanning.

7.1.3.1 Passive Scanning

In passive scanning, the device just passively scans, never transmitting anything. Passive scanning can therefore be implemented on a device that only has a receiver. By supporting only passive scanning, you can reduce the size and cost of the controller because there's no need for a transmitter. But, as mentioned earlier, depending on the amount you're producing, multi-purpose devices might end up costing less as a result of volume pricing.

7.1.3.2 Active Scanning

In active scanning, whenever a new device is discovered by the Link Layer, a scan request is sent to the advertising device, and a scan response is expected in reply. Both these scan requests and response packets are transmitted on the advertising channel. For active scanning to work efficiently, the data in the scan response must be mostly static because this data is expensive in terms of energy expended to retrieve due to the additional two packets that are transmitted or received. The data in the original advertising packet, however, can change regularly because this will always be received.

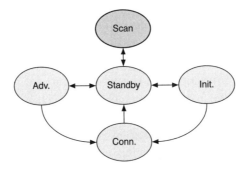

Figure 7–4 The scanning state

7.1.4 The Initiating State

To initiate a connection to another device, the Link Layer must first be placed into the initiating state. In the initiating state (see Figure 7–5), the receiver is used to listen for the device to which the initiator is attempting to connect. If an advertising packet from this device is received, the Link Layer will send a connect request to the advertiser and move into the connection state, in the assumption that the advertising device does the same. It is also possible to leave the initiating state to move back into the standby state by stopping initiating a connection.

7.1.5 The Connection State

The final state of the Link Layer state machine is the connection state (see Figure 7–6). This can be entered either via the advertising state or the initiating state. Both of these transitions are caused by an initiating device sending a connect request packet to an advertising device.

Again, this has two substates: master or slave. In the connection state, data channel packets are sent and received between the two devices. This is the only state

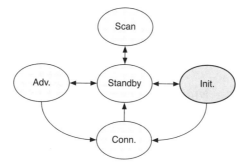

Figure 7–5 The initiating state

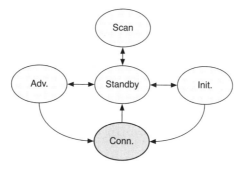

Figure 7–6 The connection state

in which the data channels are used; all other states use the advertising channels. It is only possible to leave the connection state by moving into the standby state. This is done by terminating the connection.

7.1.5.1 The Master Substate

The master connection substate can only be entered from the initiating state. A device that becomes a master must initiate the connection to the peer device. When a device is a master, it must transmit packets to the slave at regular intervals. This provides the slave with opportunities to reply and send its own data.

7.1.5.2 The Slave Substate

The slave connection substate can only be entered from the advertising state. A device that becomes a slave must have been advertising to the peer device. When a device is a slave, it cannot transmit anything until a packet from the master is received correctly. Once a packet from its master is received, the slave can transmit a packet itself. If the slave wants to transmit more data, it must wait for the master to send another packet of data back to it. Slaves can save power by just ignoring the master at any time. By doing so, the slave device can save significant quantities of power by staying "asleep."

7.1.6 Multiple State Machines

In an implementation of the Link Layer, it is possible to have multiple state machines; each state machine is separate. Using this configuration, a device can, for example, be a slave, advertise, and actively scan at the same time. Or, you could configure a device to be a master, advertise, passively scan, and initiate a connection at the same time, as illustrated in Figure 7–7. The device could also have multiple master connections to slaves at the same time.

Be aware that there are some restrictions related to deployment that are important to understand.

7.1.6.1 Not Master and Slave

The Link Layer is an "autocracy"; if a device is a master, it cannot also be a slave at the same time. Similarly, if a device is a slave, it cannot be a master at the same time. This implies that if a device is a master, it cannot advertise with a connectable advertising packet; however, it can still advertise with nonconnectable or discoverable advertising packets.

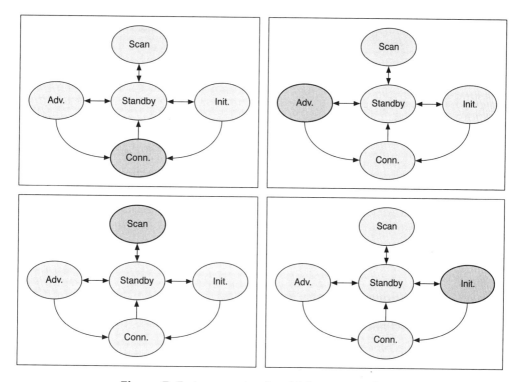

Figure 7–7 An example of multiple state machines

Consequently, if a device is already a slave, it cannot initiate a connection to another device because doing so could cause it to become the master of that device. By making this restriction, there will be no point in time when it is nondeterministic as to what the device should be doing.

The deterministic nature of the Link Layer enables Bluetooth low energy devices to be implemented using very efficient scheduling algorithms. Any nondeterministic system that is maintaining synchronization with multiple time domains will need to have very complex scheduling algorithms. These algorithms, by virtue of being nondeterministic, will also require significant processing requirements implemented in general-purpose CPUs. This does not fit in with the low-power goals for the technology. The deterministic design, therefore, allows for highly efficient algorithms that can be implemented by using discrete logic.

A device also cannot be a slave of two masters at the same time. If a device is a slave, it cannot advertise with a connectable advertising packet. Being a slave of two masters is actually a harder scenario than being a master and a slave at the time. In Bluetooth classic, this would be called *scatternet*. Bluetooth low energy does not support scatternets.

7.2 Packets

A packet is the fundamental building block of the Link Layer. A packet is really simple; it is a labeled piece of data that is transmitted by one device and received by one or more other devices. The label identifies the device that sent the data and which devices should listen to it.

Figure 7–8 shows the packet structure, which all packets share, regardless of what they're used for. At the start of the packet is a short training sequence called the preamble. After that is an access address that is used by the receiver to distinguish this packet from the background radio noise. After the access address are header and length bytes. Immediately after these is the packet's payload, followed by a cyclic redundancy check (CRC), which ensures that the payload is correct.

7.2.1 Advertising and Data Packets

In Bluetooth low energy, there are two types of packets: advertising and data packets. These packets are used for two completely different purposes. Devices use advertising packets to find and connect to other devices. Data packets are used once a connection has been made. The difference between advertising packets and data packets is that a data packet is understandable by only two devices, known as the master and slave devices; advertising packets, on the other hand, are sent by one device and can be either broadcast to any device listening or directed at a specific device.

Whether a packet is an advertising packet or a data packet is determined by the channel on which the packet is transmitted. There are 3 advertising channels and 37 data channels. If a packet is transmitted on one of the 3 advertising channels, then the packet is an advertising packet; otherwise, it is a data packet.

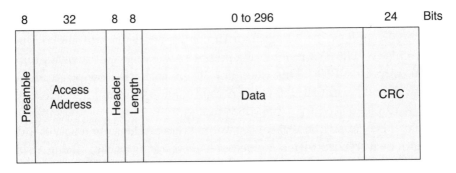

Figure 7–8 Packet structure

7.2 Packets

7.2.2 Whitening

The interesting thing about frequency-shift keying (FSK) receivers is their lack of ability to receive a very long sequence of bits of the same value. (To learn more about FSK receivers, go to Chapter 5, The Physical Layer.) For example, when transmitting a string of bits such as "000000000000", the receiver will assume that the center frequency of the transmitter has moved to the left and it will therefore lose frequency lock. It then misses the next "1" bit and fails to receive the rest of the packet. To protect against this, a *whitener* is used to randomize the packets transmitted.

A whitener is typically a very short random number generator that outputs zeros and ones in a known order for a given packet (see Table 7-1). A receiver can then use the same random number generator to recover the original bits. To keep the original information in the output sequence, the original data is combined with the random number whitener using an exclusive-or operation.

By using a random whitener combined with the original information in the packet, a string of identical bits in the original information will be converted into a sequence that is highly randomized. This reduces the chance that the receiver will lose frequency lock. If the long string of information bits were already random in nature, any further randomization will not hurt.

The whitening random number sequence is generated using a linear feedback shift register, similar to one used to calculate a CRC. The polynomial used is shown in Equation 7-1.

$$Whitener = x^7 + x^4 + x^0 \tag{7-1}$$

In this equation, x is the shift register.

The value of the shift register is initially set with the Link Layer channel number on which this packet will be transmitted, with the high bit set. This means that even if a packet is whitened on one channel that causes the receiver to lose lock, when it is transmitted on another channel, it will use a different whitening sequence, and therefore the receiver will be able to receive it. This is a very rare occurrence, but one from which the whitener allows recovery.

Table 7-1 Using a Whitener in an Exclusive-Or Operation

Input	Whitener	Output
0	0	0
0	1	1
1	0	1
1	1	0

Table 7–2 shows how the whitener is used to stop long sequences of zeros or ones from being transmitted. This example shows the first 3 bytes of data being whitened when transmitted on channel 23. This means that the binary sequence:

$$input = 00000000 : 00100000 : 00010000$$

combined with the whitener of:

$$whitener = 11110101 : 01000010 : 11011110$$

is converted to:

$$output = 11110101 : 01100010 : 11001110$$

Table 7–2 Whitener LFSR: Input Bits and Resultant Output Bits

LFSR	Input	Output
1010111	0	1
1101111	0	1
1110011	0	1
1111101	0	1
1111010	0	0
0111101	0	1
1011010	0	0
0101101	0	1
1010010	0	0
0101001	0	1
1010000	1	1
0101000	0	0
0010100	0	0
0001010	0	0
0000101	0	1
1000110	0	0
0100011	0	1
1010101	0	1
1101110	0	0
0110111	1	0
1011111	0	1
1101011	0	1
1110001	0	1
1111100	0	0

You can see that the data has become randomized with the whitener. The original input data has long sequences of single digits: 10,1,8,1,4. The output data does not have these long sequences, and critically, the longest single digit sequence is just four digits: 4,1,1,1,1,1,2,3,1,1,2,2,3,1.

7.3 Packet Structure

As shown in Figure 7–9, the packet structure is composed of a number of fields. Each of these fields is described in detail in the following subsections. Some fields contain multiple byte fields; therefore, the order of transmission of these bytes as well as the bits in these bytes also needs to be discussed.

7.3.1 Bit Order and Bytes

Packets are transmitted bit by bit, but they are composed of bytes of data. When these bytes of data are transmitted, they are transmitted with the least significant bit first. Therefore 0x80 is transmitted as 00000001, whereas 0x01 is transmitted as 10000000. Most multiple-byte fields are transmitted least significant octet first. Therefore, the value 0x010203 would be transmitted as the following:

$$11000000010000001000000$$

7.3.2 The Preamble

The first 8 bits of a packet that are transmitted are either a 01010101 or 10101010 sequence. This is a very simple alternating sequence by which a receiver can set its automatic gain control and also determine the frequencies being used for the zero and one bits.

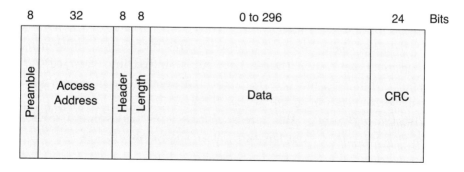

Figure 7–9 Packet structure

The reason that this sequence is very important is due to the possible range of input signal strengths with which a chip must be able to cope. The radio must be able to handle a signal at −10dBm at the antenna, all the way to −90dBm. This is a dynamic range of 80dBm. From a receiver's perspective, this means that it could receive a packet with a power of 1pW or a power of 0.1mW. An automatic gain control would therefore have to detect the input power level and adjust its gain to bring the signal into a range with which the controller can easily work.

The determination of whether the preamble is 01010101 or 10101010 is determined by the first bit of the access address that is transmitted. If the first bit of the access address is a "0", the 01010101 sequence is used. If the first bit of the access address is a "1", the 10101010 sequence is used. This always guarantees that the first 9 bits of a packet have alternating bits: either 101010101 or 010101010.

7.3.3 Access Address

The next 32 bits of a packet are the access address. This can be one of two types:

- Advertising access address
- Data access address

The advertising access address is used when broadcasting data or when advertising, scanning, or initiating connections. The data access address is used in a connection after a connection has been established between two devices.

When a controller wants to receive a packet, it always knows which access address it will be receiving. As the receiver is turned on and tuned into the correct frequency, the receiver will start to receive bits of data. Even if no other device is around transmitting at this time, the radio will pick up background radiation. Given simple probabilities of receiving pure random noise, the chance of receiving a sequence of bits that matches the preamble is fairly high; typically, once every few minutes for a low-energy device with its receiver constantly open. Therefore, the access address is used to reduce the probability of random noise causing a pseudo-packet to be received.

The Link Layer also doesn't know when the other device will be transmitting packets, so it has to keep a copy of all the possible bits that have been received for the last 40μs and check each time a new bit is shifted into this register to see if this sequence of bits now matches the expected preamble and access address. This process is called *correlation of the access address*.

For advertising channels, the access address is a fixed value: 0x8E89BED6. In binary this is transmitted from left to right as the following:

$$01101011011111011001000101110001$$

7.3 Packet Structure

This means that for an advertising packet the preamble would be 01010101. This value was chosen because it has excellent correlation properties. The fixed value means that any Bluetooth low energy device can correlate against this access address and know it is receiving an advertising packet, even though it might never have received a packet from this specific device before.

For data channels, the access address is a different random number on each and every connection between two devices. This random number, however, must adhere to a number of rules, primarily to ensure that the access address still has good whiteness.

As is explained in Section 7.2.2 on whitening, it is necessary to whiten radio transmissions to ensure that receivers can be built as easily as possible. The most basic rule is that there cannot be more than six zeros or ones anywhere in the access address. The packet also has to be different from the advertising access address by at least 1 bit. Also, the access address cannot have any repeating patterns; each octet of the access address must be different. There should be no more than 24 bit transitions, stopping the use of an alternating bit sequence. Finally, there must be at least 2 bit transitions in the last 6 bits, to ensure that just before the header starts that there are bit transitions, just in case the header whitens to a long sequence of bits.

Given the preceding rules, it can be shown that there are approximately 2^{31} possible uniquely valid random access addresses. Or in other words, it is possible to have approximately 2 billion Bluetooth low energy devices within range of one another, talking at the same time. That was probably a slight design overkill, but remember Bluetooth low energy has been designed for success. Another useful feature of this random access address for data channels is that an attacker cannot determine which two devices are in a connection by just receiving this access address. This ensures the privacy of devices during a connection.

7.3.4 Header

The next part of a packet is the header. The contents of the header depends on whether the packet is an advertising packet or a data packet.

For the advertising packet (see Figure 7–10), the header includes the advertising packet type as well as some flag bits to specify whether the packet includes public or random addresses. There are seven advertising packet types, each having a different payload format and a different behavior:

- **ADV_IND**—General advertising indication
- **ADV_DIRECT_IND**—Direct connection indication
- **ADV_NONCONN_IND**—Nonconnectable indication

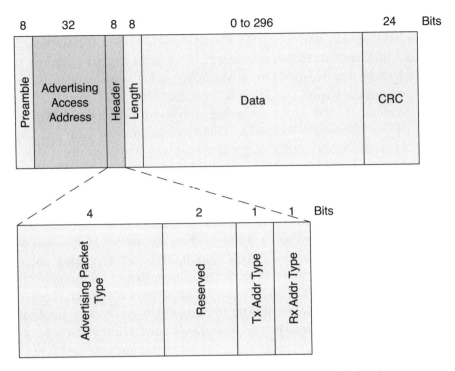

Figure 7-10 The contents of an advertising packet header

- **ADV_SCAN_IND**—Scannable indication
- **SCAN_REQ**—Active scanning request
- **SCAN_RSP**—Active scanning response
- **CONNECT_REQ**—Connection request

Figure 7-11 illustrates the header for data packets, which includes bits to enable the reliable delivery of packets, manage low power, and route the payload into either the local controller or to the host.

7.3.5 Length

For advertising packets, the length field comprises 6 bits, with valid values from 6 to 37. For data packets, it's 5 bits in length with valid values from 0 to 31. After the length field is the payload, which contains the same number of bytes of data as the value in the length field.

7.3 Packet Structure

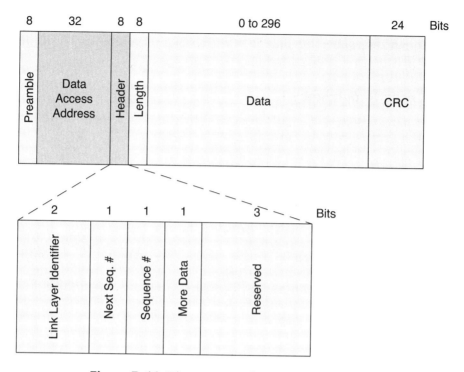

Figure 7–11 The contents of a data packet header

It might appear strange that the length field is a different length for advertising packets and data packets. The main reason for this is a design decision that accomodates 31 bytes of useful data in an advertising packet. However, an advertising packet's payload also always includes a 6-octet address for the advertising device. Adding the 6 octets of this address with the 31 octets of useful advertising data resulted in a packet length of 37 octets, and thus the requirement for a 6-bit length field.

Data packets are much easier. The size of data packets is less critical; most data being transferred is just a few octets in length, and therefore an absolutely maximal-sized packet was never considered useful. It's also interesting to note at this point that if the packet is encrypted, it includes a 4-octet message integrity check value, shortening the actual data in the payload to just 27 octets. To keep the design of the Link Layer as simple as possible, unencrypted packets are not allowed to be longer than this 27-octet limit; this reduces the complexity of buffering within the Link Layer.

7.3.6 Payload

The payload is the actual "real" data that is being transmitted. It could be advertising data about the device or service data that is being broadcast to devices in the local area. It could be additional active scan response data such as the device name and the services it implements. It could be information required to establish a connection or to maintain the connection once it is established. It could also be the application data that is being transmitted from one device to another.

7.3.7 Cyclic Redundancy Check

The last part of every packet is a 3-byte cyclic redundancy check (CRC). This CRC is calculated over the header, length, and payload fields. The CRC is a 24-bit CRC that is strong enough to detect all odd numbers of bit errors as well as 2- and 4-bit errors. This means that all 1, 2, 3, 4, 5, 7, 9, and so on, bit errors are detected in all packets.

The choice of a 24-bit CRC might appear strange, considering that most wireless standards use 16- or 32-bit CRCs. However, for the size of packet that Bluetooth low energy can send, a 32-bit CRC would not be able to detect 6-bit errors any more reliably than the 24-bit CRC used, and therefore would simply waste 8 microseconds of radio activity for every packet. If the length of the header, length, and payload fields were increased past the maximum of 39 bytes, it would be necessary to increase the CRC size to be able to detect even 4-bit errors. A 16-bit CRC, however, is not strong enough to detect the 4-bit errors over all possible 336 bits of the payload and CRC that are being protected. The 24-bit CRC is consequently the best compromise between robustness and power saving.

The polynomial used for the 24-bit CRC is as demonstrated in Equation 7-2:

$$CRC = x^{24} + x^{10} + x^9 + x^6 + x^4 + x^3 + x^1 + x^0 \quad (7\text{-}2)$$

7.4 Channels

As described in Section 5.6 of Chapter 5, The Physical Layer, Bluetooth low energy uses 40 channels. The Bluetooth low energy channels differ from classic channels because of the relaxed modulation index. This means that the radio energy for each channel is spread wider; therefore, to prevent interference between adjacent Bluetooth low-energy channels, they are separated by 2MHz, instead of the classic 1MHz.

In the Link Layer, these channels are divided into two types: advertising channels and data channels. These channel types are aligned with the advertising packets and data packets, as described earlier. When a packet is transmitted, if the packet is sent on an advertising channel, it is an advertising packet. If the packet is sent on a data channel, it is a data packet.

There are 3 advertising channels and 37 data channels, as shown in Figure 7–12 (the advertising channels are rendered in darker shading). The 3 advertising channels

7.4 Channels

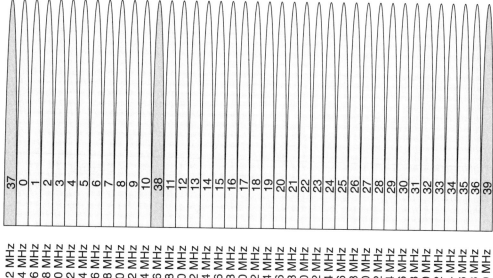

Figure 7-12 The Link Layer channel map

are not all placed in the same part of the ISM band because that would mean that any deep fade in a single part of the band would stop all advertising. Instead, the advertising channels are placed a minimum of 24MHz apart from one another.

The advertising channels are placed strategically away from significant interferers such as a Wi-Fi access point. These access points typically use one of three 802.11 channels, either channel 1, channel 6, or channel 11. These channels have center frequencies of 2412MHz, 2437MHz, and 2462MHz and a width of approximately 20MHz. This means that channel 1 extends from 2402MHz to 2422MHz, channel 6 extends from 2427MHz to 2447MHz, and channel 11 extends from 2452MHz to 2472MHz.

The advertising channels are placed at 2402MHz, 2426MHz, and 2480MHz. This means that the first advertising channel is below Wi-Fi channel 1, the second advertising channel is between Wi-Fi channel 1 and channel 6, and the third advertising channel is above Wi-Fi channel 11. This is illustrated in Figure 7-13, in which 3 Wi-Fi channels have blocked the use of data channels 0 to 8, 11 to 20, and 34 to 32. The 3 advertising channels, 37, 38, and 39, are all interference free.

The data channels are placed every 2MHz between the advertising channels. Table 7-3 shows the complete list of advertising and data channels, the Link Layer channel number, and the center frequency.

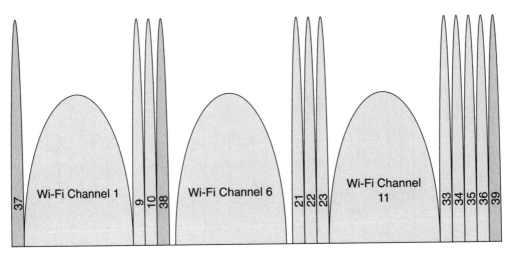

Figure 7-13 Link Layer channels and Wi-Fi channel coexistence

Table 7-3 Complete List of Advertising and Data Channels, the Link Layer Channel Number, and Center Frequency

Frequency (MHz)	LL Channel Number	Type	Frequency (MHz)	LL Channel Number	Type
2402	37	Adv	2442	18	Data
2404	0	Data	2444	19	Data
2406	1	Data	2446	20	Data
2408	2	Data	2448	21	Data
2410	3	Data	2450	22	Data
2412	4	Data	2452	23	Data
2414	5	Data	2454	24	Data
2416	6	Data	2456	25	Data
2418	7	Data	2458	26	Data
2420	8	Data	2460	27	Data
2422	9	Data	2462	28	Data
2424	10	Data	2464	29	Data
2426	38	Adv	2466	30	Data
2428	11	Data	2468	31	Data
2430	12	Data	2470	32	Data
2432	13	Data	2472	33	Data
2434	14	Data	2474	34	Data
2436	15	Data	2476	35	Data
2438	16	Data	2478	36	Data
2440	17	Data	2480	39	Adv

7.4 Channels

The advertising channels are numbered from 37 to 39; the data channels are numbered from 0 to 36. The separation of the data channel and advertising channel numbers is so that the frequency-hopping algorithm is very easy to implement.

7.4.1 Frequency Hopping

When in a data connection, a frequency-hopping algorithm is used. Because the number of data channels is 37, which is a prime number, the hopping algorithm is very simple, as demonstrated in Equation 7-3:

$$f_{n+1} = (f_n + hop) \bmod 37 \tag{7-3}$$

The *hop* value is a value that can range from 5 to 16; it is added onto the last frequency modulo 37 every time the frequency-hopping algorithm is used. This means that every frequency will be used with equal priority, regardless of the hop value. In Figure 7–14, the channels chosen, given a hop value of 13, are shown over time. Also, the algorithm can be implemented by just adding the *hop* value, comparing the value with 36, and if it is greater than this, subtracting 37. No divisions, multiplications, or other complex mathematics are required.

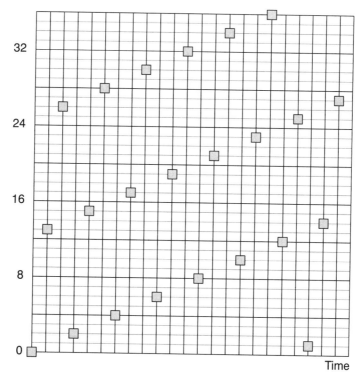

Figure 7–14 Frequency hopping of data channels over time

7.4.2 Adaptive Frequency Hopping

Adaptive frequency hopping makes it possible for a given packet to be remapped from a known bad channel to a known good channel so that the interference from other devices is reduced. To do this, a channel map of good and bad channels is kept in both devices. If the channel that would have been chosen by using Equation 7-3 is a good channel, then that channel is used; if the channel that would have been chosen is a bad channel, then it is remapped onto the set of good channels, as depicted in Figure 7–15. A minimum of two data channels must be marked as good by a master.

Suppose, for example, that a Bluetooth low energy device is in the same area as a Wi-Fi channel 1 access point that is streaming data to another Wi-Fi device. The Bluetooth low energy device would mark Link Layer data channels 0 to 8 as bad channels. This means that when the two devices are communicating, they would cycle through the channels and remap these channels to a set of good channels, as shown in Table 7–4 and Figure 7–16.

This remapping of channels ensures that even in the face of heavy interference, Bluetooth low energy will still continue to send data. It also enables the device to react very quickly to new interference. In Bluetooth classic, most controllers can react to a new interferer within just a few seconds, after which both will readily coexist without any concerns.

To assist in the remapping process, the host can inform the controller of the current channel conditions. This information could come directly from the interfering radio in the device or it could come from something much more exotic. Most Bluetooth controllers can also perform passive band scanning to determine the location and extent of interference and act on this without any input from the host.

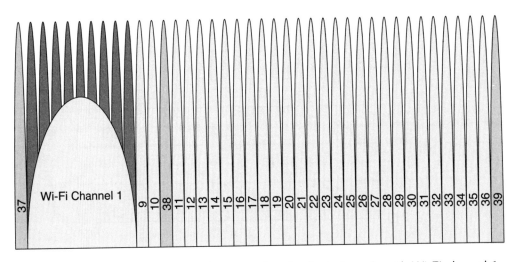

Figure 7–15 Link Layer adaptive frequency-hopping bad channels with Wi-Fi channel 1

7.4 Channels

Table 7–4 An Example of Adaptive Frequency Channel Remapping

Original Channel	Good/Bad	Remapped Channel
0	Bad	9
13	Good	13
26	Good	26
2	Bad	11
15	Good	15
28	Good	28
4	Bad	13
17	Good	17
30	Good	30
6	Bad	15
19	Good	19
32	Good	32
8	Bad	17

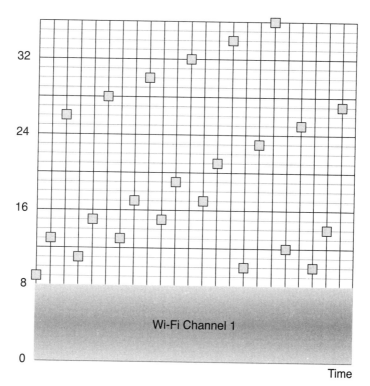

Figure 7–16 Adaptive frequency-hopping remapping

7.5 Finding Devices

A device uses the advertising channel to find another device, with one device advertising and another device scanning, as illustrated in Figure 7–17. There are four types of advertising that can be performed by devices: general, directed, nonconnectable, and discoverable.

Each time a device advertises, it transmits the same packet in each of the three advertising channels; this sequence of packets is called an *advertising event*. Apart from directed advertising, all of these advertising events can be sent as often as every 20 milliseconds to as infrequently as every 10.28 seconds. Typically, a device that is advertising would advertise once per second. The time between advertising events is called the *advertising interval*. The host can control this interval.

However, there would be a problem if devices advertised periodically because as their clocks drifted independently, two devices would constantly be advertising

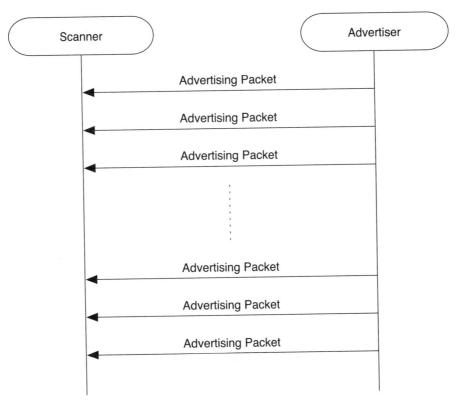

Figure 7–17 An advertiser sending advertising packets

at exactly the same time, possibly for a long period of time. To prevent this from happening, all advertising events, except directed advertising, are perturbed in time. This perturbation is done by adding a random addition time from the last advertising event of somewhere between 0 and 10 milliseconds. This means that even if two devices collide on the same advertising channel at the same time and share the same advertising interval, they will probably not collide the next time they send an advertising event.

Scanning is important to complete the picture for low-energy advertising. Scanning is required to be able to receive advertising events. How much time is available and how quickly a device needs to find another device will determine the time that will be dedicated to scanning. For example, if the user is directly touching an interface that is looking for devices, the device would scan continuously for a number of seconds, soaking up all the devices that are advertising in the area.

However, if the user is just walking around, the scanning device might only be scanning for a few milliseconds every second, or for a few hundred milliseconds every minute, looking for interesting information, depending on whether the user just arrived home, sat down in a café, or perhaps walked into a meeting room. This background scanning can then change the behavior of the device depending on where it is; if you are in the café, the phone might automatically switch to silent mode; at home, it might direct all phone calls through to the home phone system; in an office meeting room, all calls might go to your voicemail along with a message to the caller indicating that you are in a meeting and cannot be disturbed.

7.5.1 General Advertising

General advertising is the most general-purpose advertising type. A device that is generally advertising can be scanned by a scanning device or go into a connection as a slave when it receives a connect request. General advertising can be sent by a device that has no other connections; in other words, it is not a slave to another device or a master of another device.

7.5.2 Direct Advertising

Sometimes a device needs to make a connection with another device quickly. For a slave to do this, it must advertise. To allow for the fastest possible connection times, direct advertising events are used. These packets contain two addresses: the advertiser's address and the initiator's address. An initiating device that receives a direct advertising packet addressed to itself immediately sends a connect request packet in response.

These directed advertising events also have special timing requirements. The complete advertising event must be repeated every 3.75 milliseconds. This timing allows a scanning device to scan for just 3.75 milliseconds and pick up directed advertising devices.

The problem with sending packets this quickly is that the advertising channels will become congested with directed advertising packets, resulting in all other devices in the area not being able to advertise themselves. For this reason, directed advertising is not allowed to continue for more than 1.28 seconds. The controller will automatically stop the advertising if the host hasn't already done so or if a connection has not been established. Once the 1.28 seconds have expired, the host would then just be able to use general advertising at a much lower duty cycle to allow the device to still be connectable.

When using directed advertising, a device cannot be actively scanned. Also, directed advertising packets cannot have any additional data in the payload of the packet; they contain only the two addresses needed, and nothing more.

7.5.3 Nonconnectable Advertising

Devices that don't want to be connectable use nonconnectable advertising events. Typical uses of this include devices that are broadcasting data and have no intention of being either scannable or connectable. This is the only type of advertising that a device equipped with only a transmitter can use.

A nonconnectable advertising device will never enter the connection state; therefore, it can only transition between the advertising state and the standby state when asked to do so by the host.

7.5.4 Discoverable Advertising

The final type of advertising event is the discoverable advertising event. This cannot be used to initiate a connection, but it can be used to allow another device to scan the advertising device. This means that the device is discoverable, both for advertising data and scan response data, but cannot be connectable. This is an advanced form of broadcast data, whereby the dynamic data can be included in the advertising data, whereas static data would be included in the scan response data.

Discoverable advertising will never enter the connection state; instead, it moves back to the standby state when it is stopped.

7.6 Broadcasting

As explained in the previous section, devices can advertise. However, for a device to be considered a broadcasting device, it must also include some useful data in that advertisement. This means that you can broadcast with three of the four advertising events: general advertising, nonconnectable advertising, and discoverable advertising.

When broadcasting, the data is labeled within the advertising packets. This is done because not all devices will understand all possible broadcast data. As such, there needs to be a way for the broadcast data to be both labeled and sized. Each piece of data starts with a length field that indicates the length of the following type and data fields. Next is a type field that a receiver will use to determine if it understands the following data (see Chapter 12, The Generic Access Profile, Section 12.5). By using this "length : type : data" format, devices that do not understand a particular type of data can skip over it because they know the size of the data and can therefore continue with the next piece of data.

Broadcast data can be received by any passive or active scanning devices nearby. Broadcast data cannot be acknowledged. A broadcasting device also doesn't know if any device received its data or if any device is attempting to listen to the data. Therefore, broadcasting must be considered to be an unreliable operation.

7.7 Creating Connections

If the data transfers are more complex than can be performed by broadcasting the data, or the data needs to be reliably delivered to another device, a connection will be required. A connection uses the data channels to reliably send information, in two directions, between two devices. It uses adaptive frequency hopping to be robust and a very low duty cycle to keep the power consumption as small as possible.

As illustrated in Figure 7–18, the first step in creating a connection is for one device to advertise by using a connectable advertising event and for another device to initiate a connection to the advertising device. To make a connection, either the general advertising event or the direct advertising event types must be transmitted by the advertiser. When the initiator receives the advertising packet from the correct device, it sends a connect request back to the advertiser. This connection request packet includes everything that is needed at the start of the connection, which is presented in the following list:

- Access Address to be used in the connection
- CRC initialization value
- Transmit window size
- Transmit window offset

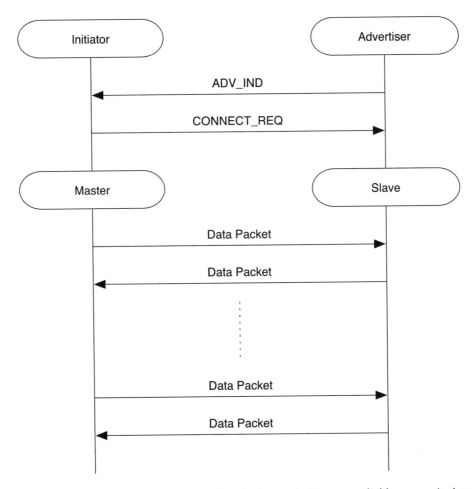

Figure 7–18 Creating connections with which two devices can reliably transmit data

- Connection interval
- Slave latency
- Supervision timeout
- Adaptive frequency-hopping channel map
- Frequency-hop algorithm increment
- Sleep clock accuracy

Once the connect request packet is sent or received, the devices are connected and data packets can be exchanged.

7.7 Creating Connections

7.7.1 Access Address

The master always determines the access address that will be used in the connection. The value is random, adhering to a few rules, as detailed in Section 7.3.3. If the master has multiple slaves, it will choose a different random access address for each slave. The randomness of this value ensures that the probability for collisions between different masters and slaves is very low. The randomness also enhances privacy by not allowing a scanner to determine which two devices are communicating.

7.7.2 CRC Initialization

The CRC initialization value is another random value chosen by the master. This is random because a small probability exists that two masters in the same area could use the same access address to talk to different slaves. If this did occur, the slaves could receive interfering data from the wrong master. By randomizing the CRC initialization value for each slave, the probability of having two masters and slaves with the same access address and the same CRC initialization value is very small.

7.7.3 Transmit Window

Advertising is always done based on the timing of the slave. The slave is the device that needs to save the most power, so this is the correct design decision. However, if the master device is already doing something else, possibly something more important, it must interleave the Bluetooth low energy activity around its other traffic. During connection setup, this information is conveyed in two parameters: window size and window interval.

The transmit window starts after the end of the connection request packet, plus an additional mandatory delay of 1.25 milliseconds, plus the transmit window offset. The transmit window offset can have any value from 0 to the connection interval, in multiples of 1.25 milliseconds. At the start of the transmit window, the slave device opens up its receiver and waits for a packet to come from the master. If this packet is not received within transmit window size, the slave aborts and tries again one connection interval later.

The most interesting thing about the connection process is that once the connection request has been transmitted, the master believes it has a connection; the connection has been created but not proven to have been established. Once the slave receives the connection request, it believes it is also in a connection; again, the connection has been created but not proven to have been established.

In the interests of efficiency, the host is notified immediately when the connection is created. The connection might not succeed, the slave might not receive the connection request, or the two devices might be such a long distance apart that the connection has a high probability of failure. However, because the host is notified of the connection being created, it can start to send data down into the controller, ready for it to send in the very first packet over the air, saving time and therefore energy. Because the first data packet is not sent until after the mandatory 1.25-millisecond delay that follows the creation of the connection and the host stack notification, the host stack should have sufficient time to provide this data into the controller, using the very first opportunity to send data. This mandatory delay also provides any batteries with time to recover from the possibly exhausting advertising procedure before the connection is created.

The connection is only considered established once a packet has been acknowledged. Establishment doesn't change how the connection works, but it does change the link supervision timeout, from just six connection timeouts, to the value in the connection request message. This ensures that if the connection is not established quickly, it is terminated immediately.

7.7.4 Connection Events

When in a connection, the master has to transmit a packet to the slave once every connection event. A connection event is the start of a group of data packets that are sent from the master to the slave and back again. A connection event is always conducted on a single frequency, with each packet transmitted 150 microseconds after the end of the last packet.

The connection interval determines how frequently the master will talk to the slave; it is the time between the start of the last connection event and the start of the next connection event. This can be any period from 7.5 milliseconds to 4 seconds in multiples of 1.25 milliseconds. To determine how infrequently the slave is allowed to talk to the master, the slave latency is used. This is a multiple of the connection interval and therefore determines how many times the slave can ignore the master before it is forced to listen. It should be noted, however, that this must still be quicker than the supervision timeout.

As illustrated in Figure 7–19, each connection event is a connection interval apart. Each connection event starts with a single packet from the master, and can continue until either the master or slave device stops responding. The times between connection events contain no packets from the master to this slave or the other way around.

For example, if the connection interval is 100 milliseconds and the slave latency is 9, the slave can ignore the master for 9 connection events but is forced to listen for the 10th, or once every second. The supervision timeout, therefore, must be a minimum of 1010 milliseconds. At the extreme end, the maximum supervision timeout is 32 seconds, and, therefore, with a connection interval of 100 milliseconds, the slave latency must be 319 or less.

7.7 Creating Connections

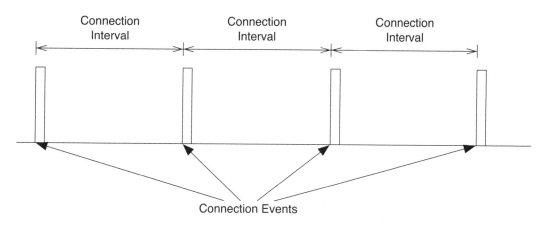

Figure 7–19 Connection events

However, it is not a good idea to give the slave just one opportunity to listen for the master within the supervision timeout when the slave is using its slave latency to the maximum. Thus, it is recommended that the slave is given six opportunities to listen. Therefore, in the preceding example, if the connection interval is 100 milliseconds and the slave latency is 9, the supervision timeout should be a minimum of 6 seconds, allowing the slave to listen a minimum of six times before the link is dropped.

7.7.5 Channel Map

The adaptive frequency-hopping channel map is a bit mask of the data channels that are known to be good or bad. Because there are 37 data channels, the channel map is 37 bits in length. If the bit is set to one, the channel is a good channel and will be used for data traffic. If the bit is zero, the channel is a bad channel and will never be used for data traffic.

The frequency-hopping algorithm's hop value is a random number between 5 and 16. It is used in the frequency-hopping algorithm before the adaptive remapping algorithm is used, as described in the frequency-hopping Section 7.4.1. The number zero is obviously not allowed because this could mean that the frequency would never change.

Very low hop numbers are not desirable because most interferers are typically more than a couple of megahertz in width; therefore, having a very small number would not move the next transmission opportunity away from the interferer quickly enough, causing continued interference. The same logic is used for values 17 and higher. With a hop increment of 17, for example, every other frequency used would be just 3 channels away because of the modulo 37 operations used in the frequency-hopping algorithm (refer to Equation 7-3).

7.7.6 Sleep Clock Accuracy

Finally, the sleep clock accuracy value is sent from the master to the slave. This value determines the range of accuracies that the clock is able to guarantee. If the clock is timed from a crystal, the crystal will have a known accuracy over the temperature range, from, for example, 20ppm at room temperature to 50ppm at 0°C or 85°C. Therefore, the clock accuracy would have to state that this device has a clock accuracy of up to 50ppm.

The clock accuracy is used to determine the uncertainty window of the slave device at a connection event. If the slave has not synchronized its timing with that of the master for 1 second, and both devices have a timing accuracy of 500ppm, then the combined uncertainty of 1,000ppm has to be multiplied with the time away, to give a 1 millisecond uncertainty window. This means that the slave must wake up 1 millisecond early and stay on for an extra 1 millisecond, just in case the master and slave clocks have both drifted at the maximum ppm in different directions.

Having more accurate clocks can help save power. For example, if the crystals used in two devices were, for instance, 150ppm in one device and 50ppm in another, then the combined accuracy would be just 200ppm. So, after one second away, the slave would have to wake up just 200 microseconds early and stay on for an extra 200 microseconds. If a device is waking up infrequently, this could run for 5 times longer than the two devices with 500ppm crystals. It is therefore recommended that high-accuracy crystals be used in devices that have very low-power requirements and need to maintain a connection for some time.

7.8 Sending Data

Once in a connection, devices can send data to one another. This is done by sending data packets at connection events. Data packets are distinct from advertising packets because they are private communications between two devices rather than broadcast communications to any device that is listening. The biggest differences between advertising packets and data packets are the length of the payload that is possible and the packet header.

The length of the payload in a data packet can be anywhere from 0 octets in length to 31 octets. A zero-length payload in a data packet is an empty packet; it has no application data but can still include some information from the packet header. The maximum length payload (at 31 octets) is smaller than the advertising packet's maximum length. It is also only possible to have this large a packet if it is encrypted.

7.8 Sending Data

An unencrypted data packet has a maximum of just 27 octets of data in it to allow for the retransmission of a data packet, even after encryption has been established between the data packet being sent into the controller and encryption being enabled in the controller.

7.8.1 Data Header

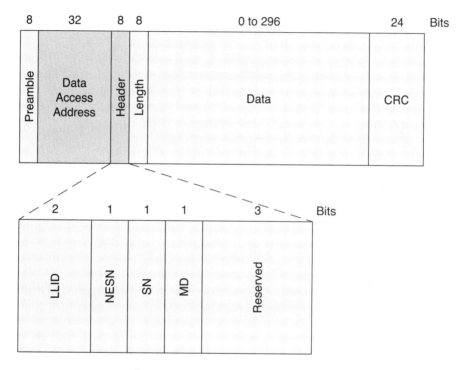

Figure 7–20 The data packet header

The data packet header, as shown in Figure 7–20, contains just the following four fields:

- Logical link identifier (LLID)
- Sequence number (SN)
- Next expected sequence number (NESN)
- More data (MD)

There is no "packet is encrypted" bit because this is a modal property of the connection, just like for the adaptive frequency hopping or the connection event intervals.

7.8.2 Logical Link Identifier

The logical link identifier (LLID) is used to determine if the packet contains one of the following types of data:

- Link Layer control packet (11)—This is used for managing the connection
- Start of a higher-layer packet (10)—Or for a complete packet
- Continuation of a higher-layer packet (01)

If the packet is a Link Layer control packet, this is indicated by the logical link identifier, and this data is passed directly to the Link Layer control entity. The meaning of the data within this packet is therefore determined by the Link Layer control entity, as described in Section 7.10.

All other packets are to or from the host. The host is able to send packets larger than the maximum 27 octets of data that can be included in a single Link Layer data packet; therefore, it is able to segment these. To do this, the packet is labeled as either a start of a higher layer data packet or a continuation of a higher layer data packet. This is illustrated in Figure 7–21, in which a very long higher layer data packet is split over three Link Layer data packets. The first data packet is labeled with an LLID of *"start"*, whereas the other two data packets are labeled with an LLID of *"continuation"*.

This is interesting from two standpoints. First, the Link Layer doesn't require knowledge of the ultimate size of the packet at the start of the packet. It is possible to send *start, continuation, continuation, ... continuation, continuation*, before sending another start message. The number of continuation messages is not fixed at the start of the message.

This allows the second interesting side effect: it is possible to always send zero-length continuation messages without any impact on the higher layer data. This allows empty packets to always be sent, meaning that simple acknowledgement

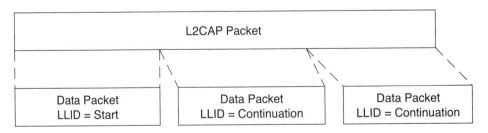

Figure 7–21 Packet fragmentation

messages can always be sent as zero-length continuation messages. These zero-length continuation packets are known as empty packets.

7.8.3 Sequence Numbers

To enable the reliable transfer of data, all data packets have sequence numbers. The sequence number for each new data packet sent is different from the last data packet's sequence number, with the first packet in a connection having a sequence number of zero. This allows a receiving device to determine if the next packet that is received is a retransmission of the previous packet because the sequence number is the same or a transmission of a new packet because the sequence number is different.

In data packets, there is a single bit for the sequence number, starting at zero for the first data packet that is sent. The sequence number then alternates between one and zero for each new data packet that is sent by a device.

7.8.4 Acknowledgement

To perform acknowledgement of a data packet, a single bit is used. This is called the *next expected sequence number bit*. This informs the receiving device of the next sequence number that the transmitting device is expecting it to send.

If the packet received by a device has the sequence number zero, the next expected sequence number that it receives must be one; otherwise, the packet would have been retransmitted. Therefore, it's possible to signal if the packet received was received correctly or if the packet needs to be retransmitted. This is illustrated in Figure 7–22.

7.8.5 More Data

The final bit in the data channel packet header is the more data bit. This signals to the peer device that the transmitting device has more data ready to be sent. The peer receiving device upon seeing the more data bit set in a received packet should continue to communicate in this connection event. This automatically extends connection events while there is still data to be sent. It also quickly closes connection events for which there is no more data to be sent. The more data bit also allows a device that needs to save power to close the connection event gracefully and quickly by setting its more data bit to zero. The more data bit can therefore be used to enable lots of data to be reliably delivered in a very efficient manner by using as few transmitted packets as possible.

7.8.6 Examples of the Use of Sequence Numbers and More Data

The processing of sequence numbers, next expected sequence numbers, and more data bits is shown in Figure 7–22 and described in the following:

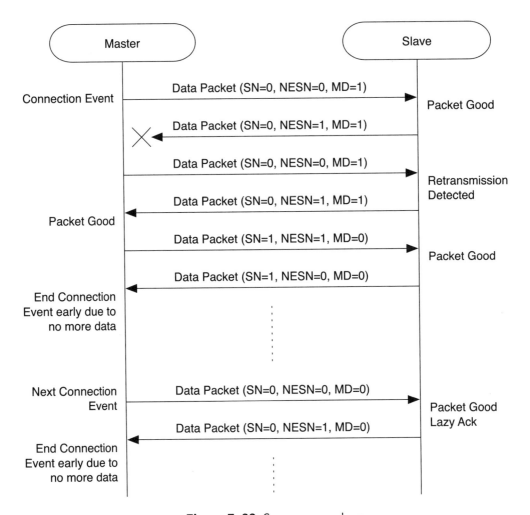

Figure 7-22 Sequence numbers

1. The master transmits its first packet by using the default sequence number of zero and the next expected sequence number of zero. The master also sets the more data bit because it has two packets of data to send ($SN_{master}=0$, $NESN_{master}=0$, $MD_{master}=1$). This packet was received correctly by the slave; therefore, the slave's next expected sequence number is updated ($NESN_{slave}=1$).

2. The slave's first packet ($SN_{slave}=0$) was transmitted with the newly updated next expected sequence number ($NESN_{slave}=1$). This packet also has the more data bit set ($MD_{slave}=1$) because the slave is about to send some interesting data. This packet was not received by the master, so the master's next expected

7.8 Sending Data

sequence number was not changed. The slave continues to listen to the master because the more data bits for both the master and the slave were set.

3. The master's second transmission ($SN_{master}=0$, $NESN_{master}=0$, $MD_{master}=1$) is a retransmission of its first packet. The master never received a packet from the slave and therefore must retransmit its packet. The slave receives this packet and detects that this is a retransmission of the last packet it received because the sequence number is identical; thus, the slave does not update its next expected sequence number. The slave sees that the master has more data to send to it, so it continues this connection event by transmitting another packet.

4. The slave retransmits its first packet ($SN_{slave}=0$, $NESN_{slave}=1$, $MD_{slave}=1$). The master receives this packet successfully and updates its next expected sequence number ($NESN_{master}=1$). The master sees that the slave has more data to send and continues this connection event by transmitting another packet.

5. The master's third transmission is a new packet requiring a new sequence number ($SN_{master}=1$, $NESN_{master}=1$, $MD_{master}=0$). This packet contains the last of the data the master has to send, so the master has set the more data bit to zero. The slave receives this packet successfully and updates its next expected sequence number ($NESN_{slave}=0$). The slave still has more data it wants to send, so it continues the connection event.

6. The slave's third transmission is a new packet requiring a new sequence number ($SN_{slave}=1$, $NESN_{slave}=0$, $MD_{slave}=0$). The slave's more data bit has been set to zero to indicate that it doesn't have any more data to send. The master receives this packet correctly and updates the next expected sequence number ($NESN_{master}=0$). Because the last packets of the master and slave both indicated that neither device has any more data, the connection event is immediately closed.

7. Some time later, the master wakes up at the next connection event and transmits a new packet to the slave with a new sequence number and the latest next expected sequence number ($SN_{master}=0$, $NESN_{master}=0$, $MD_{master}=0$). This packet also lazily acknowledges the last packet from the slave. The slave receives this packet successfully and updates its next expected sequence number ($NESN_{slave}=1$). The slave has no more data to send but needs to respond with an empty packet.

8. The slave's fourth transmission is an empty packet using a new sequence number ($SN_{slave}=0$, $NESN_{slave}=1$, $MD_{slave}=0$). The slave also has no more data to send; therefore, it sets its mode data bit to zero. The master receives

this packet successfully and updates its next expected sequence number ($NESN_{master}=1$). Because the last packets of the master and slave both indicated that neither device has any more data, the connection event is immediately closed.

As the preceding example illustrates, the sequence number and next expected sequence number are always in lock-step with one another. This ensures that packets are always reliably delivered and in order. A packet is not considered to be received when the CRC fails to verify that the contents of the packet have been received correctly.

It is also possible to enable flow control by using the next expected sequence number. If a device does not have enough buffer space to process a message at a given time, it is not able to update the next expected sequence number. This forces the other device to retransmit this message, effectively pushing the buffering requirements onto the sending device and away from the receiving device.

There is one other effect of this whole process on connection events. If a packet is not successfully received because a few bit errors cause the CRC to fail, and this happens again on the same packet within a single connection event, the two devices will stop using that connection event. The devices will then resynchronize at the next connection event and try again. This means that if a given channel is being blocked due to interference, then very quickly the two devices will discover this interference and stop using that channel. By moving to a new channel quickly, the interference caused by transmitting on the blocked channel is mitigated almost immediately, but data is still delivered to the other device very quickly at the next connection event.

7.9 Encryption

When in a connection, data within the payload can be encrypted. This encryption can ensure confidentiality of the data against attackers. Confidentiality means that a third party cannot intercept the messages, decipher them, and read the original contents of the messages because the "attacker" does not have the shared secret used to encrypt the link.

Encrypted packets also include a message integrity check value that ensures the data is authenticated. Authentication means that the validity of the sender can be confirmed by calculating a signature of the encrypted data with a shared secret. This prevents a third party from changing any of the bits in the packet. Authentication allows the receiver of a message to know that the data packet it has just received was sent by a device it trusts. A personal identification number (PIN) used to authenticate bank card holders is a classic example of authentication; the PIN verifies that the authorized person is using the bank card.

Encrypted packets also include a packet counter to stop replay attacks. A replay attack is one in which an attacker intercepts a given message and then at a later date replays this message in the hope that it will result in a response. For example, without replay attack protection, it could be possible to scan for lots of packets being transmitted by a device and then replay these packets and see what happens. If the receiving device was a sewage valve near a city park, the results could be "interesting." Clearly, protecting against replay attacks is something very important.

7.9.1 AES

All encryption and authentication in Bluetooth low energy is built around a single encryption engine called the Advanced Encryption System (AES). This encryption engine was originally designed as part of a United States government program to find a suitable encryption engine for the future. It has since been adopted by many wired and wireless standards and has so far held up well against the attempts by security researchers to find weaknesses in its algorithms.

AES can be built in multiple forms, typically determined by the size of the blocks of data and keys that it can process at any given time. In Bluetooth low energy, the 128-bit key size and 128-bit data blocks are used. This means that all keys are 128 bits in length, and up to 16 octets of encrypted data can be created at a time.

The AES encryption block is very simple: it takes two inputs and generates one output. The two inputs are a 128-bit key value and a 128-bit block of plain-text data, and the output is a 128-bit block of encrypted data. The reason the two inputs are labeled as key and plain-text data is that the key must be processed first before being used in the encryption block, whereas the plain text is immediately processed by the encryption block. Therefore, it is more efficient to set up the key once and then pass in different plain-text blocks and encrypt them quickly than it is to use a different key for each block.

Thus, the encryption of some data, *plaintext*, with a key, *key*, using an algorithm, E, to produce encrypted text, *ciphertext*, can be represented by using the function in Equation 7-4:

$$ciphertext = E_{key}(plaintext) \quad (7\text{-}4)$$

In Bluetooth low energy, the AES encryption engine is used for four basic functions:

- Encrypting payload data
- Calculating a message integrity check value
- Signing data
- Generating private addresses

The signing of data is defined by the Security Manager, and the generating of private addresses is defined by the Generic Access Profile.

7.9.2 Encrypting Payload Data

To encrypt the payload data, the payload is split into 16-byte blocks, and for each block a cipher bit stream is generated. This cipher bit stream is then Exclusive-Or'ed with the plain text. This is defined by using the standard IETF RFC 3610. This standard defines a method for encryption and authentication using Counter with Cipher Block Chaining-Message Authentication Code Mode or CCM.[1] This is a standard encryption algorithm for any size key and any size message.

In Bluetooth low energy, the A_x encryption blocks are used. These are initialized plain-text blocks that have a known format and include a *nonce* composed of a packet counter, a direction bit, and an initialization vector (IV). In the equation that follows, the || notation means concatenation.

$$nonce = Packet\,Counter\,||\,Direction\,||\,IV$$

The packet counter is a 39-bit value that is incremented for each new non-empty packet that is transmitted. The packet counter always starts with zero when encryption is enabled. Empty packets are not encrypted; therefore, they do not need to increment the packet counter. The initialization vector is a random value that is 64 bits in length, where 32 bits of this vector are contributed by each device in the encrypted link (see Section 7.10.3). Therefore, the *nonce* is 13 bytes in length.

The other octets of the A_x are per the CCM specification. The first octet is a flags field that will always be set to 0x01 to indicate that this is an A_x block. The last two octets are the block counter. The block counter is set to 0x0001 when used to encrypt the first 16 octets of the payload (C_{Block1}) and to 0x0002 when used to encrypt the second 11 octets of the payload (C_{Block2}). The CCM specification is also used to encrypt the message integrity check value (M_{MIC}), and for this, the block counter is set to 0x0000.

$$C_{MIC} = E_{key}(0x01\,||\,nonce\,||\,0x0000)$$

$$C_{Block1} = E_{key}(0x01\,||\,nonce\,||\,0x0001)$$

$$C_{Block2} = E_{key}(0x01\,||\,nonce\,||\,0x0002)$$

---

1. Cipher Block Chaining = CBC; Message Authentication Code = MAC; Counter with CBC-MAC = CCM.

7.9 Encryption

The cipher blocks are then Exclusive-Or'd with the various parts of the message to generate the *encrypted* payload.

$$encrypted = C_{Block1} \oplus Block1 \,||\, C_{Block2} \oplus Block2 \,||\, C_{MIC} \oplus MIC$$

This encrypted payload is then sent to the peer device. Because the peer device knows the shared secret—the *key* value—it can decrypt the message by using the same packet counter, direction, and IV values. When the encrypted payload is sent, the CRC is calculated over the encrypted payload, not the original payload blocks. The header and length fields of the data packet are never encrypted.

7.9.3 Message Integrity Check

The message integrity check (MIC) value is used to authenticate the sender of the data packet, and this MIC is inserted between the Data and the CRC, as illustrated in Figure 7–23. This ensures that the encrypted packet is sent by the peer device and not by a third-party attacker. To calculate the MIC, the AES encryption engine is used again. This time, the output of one block is used as the input to the next block, chaining together the blocks to ensure that every bit in the original message is as important as every other bit in calculating the MIC.

To calculate the MIC, the same *nonce* is used for encrypting the payload. Three or four B blocks are used. The first B_0 block contains the nonce and the original length of the data being authenticated. This length field comprises 16 bits, even though the maximum size of the payload that can be authenticated in low energy is just 27 octets.

$$B_0 = 0x49 \,||\, nonce \,||\, length$$

The next B_1 block contains additional data that should be authenticated with the payload but is not contained within the payload. In Bluetooth low energy, this is used to authenticate some of the bits in the header of the packet. The only bits

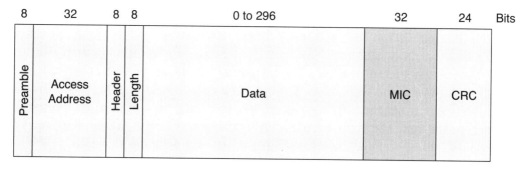

Figure 7–23 The encrypted data format

that need to be authenticated are the logical link identifier bits. All other bits in the header are masked to zero; this simplifies calculation and allows precalculation of blocks without having to know values such as SN or NESN, and so on because these bits are not important from a security point of view.

$$B_1 = 0x0001 \,||\, header_{masked} \,||\, 0x00000000000000000000000000$$

The next block or two contains the actual payload data being authenticated. B_2 contains the payload data from octets 0 to 15. B_3 contains the payload data from octets 16 to 26.

To calculate the MIC, these blocks are then chained together by using a single *key* as used for encrypting the payload. Only the most significant 32 bits of the payload are used in the packet.

$$X_0 = E_{key}(B_0)$$

$$X_1 = E_{key}(X_0 \oplus B_1)$$

$$X_2 = E_{key}(X_1 \oplus B_2) \text{ where } B_2 = Payload_{[0..127]}$$

$$MIC = E_{key}(X_2 \oplus B_3)_{[128..96]} \text{ where } B_3 = Payload_{[128..215]}$$

When an encrypted packet is received, the same MIC calculation is performed on the receiving device to check that the MIC value computes to the same, given the same inputs. If the value does not compute correctly, the connection is immediately disconnected. No further communication will occur, and the peer device will automatically eventually enter supervision timeout. This appears at first glance to be a very drastic approach; however, the MIC will not even be checked if the CRC fails—the packet will just be rejected and a new packet retransmitted from the peer.

The only way that an MIC can fail is if an attacker is currently attempting to attack the link or if a number of bit errors were falsely accepted by the CRC, causing the MIC to fail. In the first case, the safest approach is to immediately disconnect the link because it might already be compromised. In the second case, the data contained in the packet is already compromised because the CRC has falsely identified it as a correct packet. So again, the safest approach is to assume the worst and disconnect the link. Once the connection has been dropped, the two devices can quickly reconnect, establish a new initialization vector, and reencrypt the link. The approach of reestablishing a new initialization vector refreshes the nonce and, therefore, the encryption that is used on the connection.

It is also possible to reestablish a new initialization vector while in a connection, if needed. Given that the packet counter has a fixed size, and that the encryption is only considered secure if nonce values are never repeated, it is necessary to refresh the initialization vector periodically. This will not occur very frequently; there are a total of $2^{39} - 1$ packets that can be sent in a connection before the nonce would repeat, which would take over 12 years of continuously transmitting packets to get anywhere close to wrapping. However, Bluetooth low energy has been designed for success and for connections to be stable for many years. As such, it is already a defined process to refresh the nonce. To do this, a new initialization vector is generated by using the encryption pause and resume procedures, as defined in Section 7.10.4.

7.10 Managing Connections

Once two devices are in a connection, they can send and receive data and manage the connection. Connection management involves sending Link Layer control messages. There are only seven Link Layer control procedures:

- Updating the connection parameters
- Changing the adaptive frequency-hopping channel map
- Encrypting the link
- Reencrypting the link
- Exchanging feature bits
- Exchanging version information
- Terminating the link

7.10.1 Connection Parameter Update

When the connection is created, the connection parameters are sent in the connection request packet, as detailed in Section 7.7. After a connection has been active for a period of time, the connection parameters might no longer be suitable for the services being used over this connection. The connection parameters will need to be updated for the services to be used efficiently. Instead of disconnecting the link and then reconnecting it with different connection parameters, it is possible to do a connection parameter update within the link, as illustrated in Figure 7–24.

To do this, the master sends a connection update request to the slave with the new parameters by using LL_CONNECTION_UPDATE_REQ. There is no negotiation of these parameters; the slave must use them. If the slave doesn't accept the parameters being suggested, it only has one option available to it: disconnect the

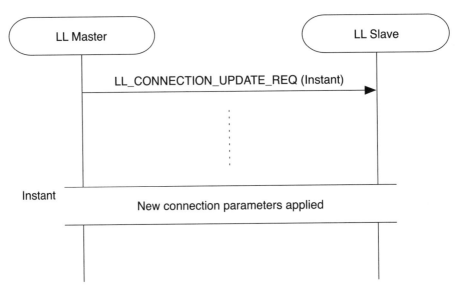

Figure 7–24 Performing a connection update procedure

link. The connection update request includes a subset of parameters that were used in the connection request message sent earlier during connection creation and one additional parameter, called the instant:

- Transmit window size
- Transmit window offset
- Connection interval
- Slave latency
- Supervision timeout
- Instant

The instant is the parameter that determines from when the connection update will start. When the master sends the message, it picks a time in the future when the connection update will be actioned and includes it in the message. The slave, upon receiving the message, will remember this instant future time and then wait until the specified time before moving to the new connection parameters. This helps solve one of the largest problems in wireless systems: packet retransmission. As long as the packet is retransmitted enough times and eventually gets through before the instant passes, the procedure will work well. If, however, the packet does not get through in time, the link will probably drop.

Given that Bluetooth low energy has no clock, the only way to determine an instant is to count connection events. Therefore, each connection event is counted, with zero being the first connection event in the link; the one that was transmitted in the first transmit window after the connection request. The instant, therefore, is the connection event count at which the new parameters will be used. The master should provide enough opportunity for the slave to receive this packet. Even at maximum latency, this should typically allow at least six attempts for the message to be sent by the master to the slave. If the slave latency is 500 milliseconds, then the instant would typically be placed at least 3 seconds in the future.

Once the instant arrives, the slave listens for the transmit window, just like during connection creation. This allows the master to shift the timing of the slaves, both within the 1.25-millisecond slot but also at the gross level. This better allows the master device that is also a Bluetooth classic device to align its own Bluetooth low energy slaves to those of its other activities. Once this procedure is complete, a new connection interval, supervision timeout, and slave latency values are used.

7.10.2 Adaptive Frequency Hopping

Adaptive frequency hopping is very important for the successful survival of any radio technology in an open wireless band. Unfortunately, some technologies don't perform adaptive frequency hopping and are therefore susceptible to interference. The biggest problem with adaptive frequency hopping, especially in model devices, is that the set of channels at any given time that can be considered good or bad can be considered to be constantly changing. This means that there needs to be signaling to allow devices to change this channel map. This procedure is shown in Figure 7–25.

The adaptive frequency-hopping updates are sent in a channel map request packet LL_CHANNEL_MAP_REQ. This is sent from the master to the slave and includes only the following two parameters:

- New channel map
- Instant

The instant is the same concept as the instant used in the connection update. It determines a point in time that the new channel map will be used. At the instant, and afterward, the new channel map is used for all connection events in the future at least until the next time the channel map is updated.

The channel map is a 37-bit field that has one bit for each data channel. If a given channel's bit is set to one, the channel is considered good and will be used; if a given channel's bit is set to zero, the channel is considered bad and will not be used.

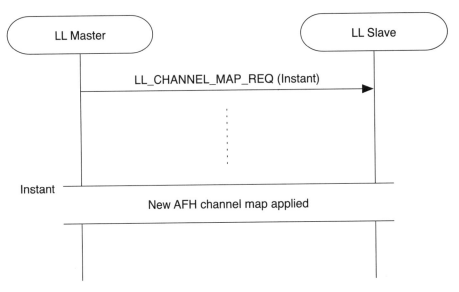

Figure 7–25 The Channel map update procedure

The channel update request can only be sent again after the instant has passed. This places a restriction on how fast the connection's channel map can be updated. Typically, the channel map would be updated only when the connection is performing poorly using its current set of channels or when the host determines that the current bad channels are now good again. No Link Layer control procedures are provided to allow a slave device to change the channel map or even notify its master of its own channel conditions.

7.10.3 Starting Encryption

To start encryption, the link must be unencrypted. To encrypt the link, both the nonce and a session key (SK) need to be created. The nonce requires 4 octets of information to be contributed by each device, and the session key requires 8 octets of information to be contributed by each device. An additional key is also required, called the long-term key (LTK). This is the shared secret that is established during pairing (for more information, go to Chapter 11, Security, Section 11.2).

To start encryption, as illustrated in Figure 7–26, the master first transmits an encryption request message (LL_ENC_REQ) to the slave. The slave then responds with an encryption response message (LL_ENC_RSP). The encryption request packet from the master includes its 4-byte contribution to the initialization vector, 8 bytes of session key diversifier, and some additional information that the slave transmitted to it when they initially paired. This additional information is static for a given master, and the slave can use this information to determine with

7.10 Managing Connections

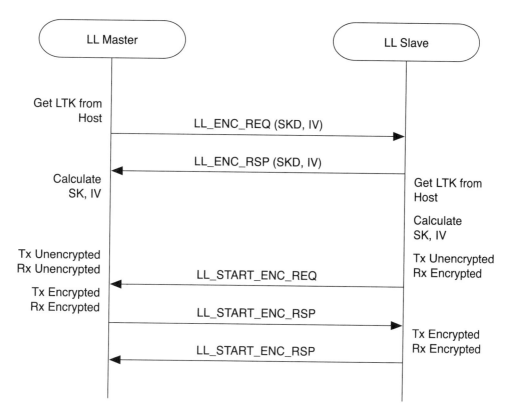

Figure 7–26 Starting the encryption procedure

which master it is communicating and possibly derive the LTK for the master from this information. By doing this, the slave might not need to store any information about bonded devices. The encryption response packet from the slave includes its 4-byte contribution to the initialization vector and its contribution to the session key diversifier.

If the LTK is not available on the slave side, the slave will then immediately send a reject indication to the master along with the reason it rejected the encryption. If the LTK is available, the slave will start a three-way handshake to begin encryption. The three-way handshake is required because the slave, for example, must be able to transmit an unencrypted packet to the master, but it must be able to receive an encrypted packet back. Thus, this handshake procedure moves the two devices in lockstep into a fully encrypted link.

When encryption is started, a session key is used to encrypt the link. The session key is calculated from the LTK and the session key diversifier contributed to by both devices.

The session key diversifier enables an LTK to be used multiple times. This is done by ensuring that each time a connection is encrypted a different encryption key is derived from the session key. The master and slave contribute half of this diversifier to ensure that even if one device is an attacker, the other device can force a different diversifier and therefore a different encryption key. The primary reason all this is done is to protect against the single weakness in AES: A key cannot be used more than once, ever. Therefore, even though we have a shared secret, LTK, we cannot use this to encrypt the application data. Instead, we must diversify this LTK into a session key, SK.

The two session key diversifiers (SKD), SKD_{master} contributed by the master, and SKD_{slave} contributed by the slave, are concatenated together and used as the *plaintext* input into the AES encryption engine (refer to Equation 7-4). The LTK is used as the *key* input into the AES encryption engine. The output of the AES encryption engine is the session key that is used as the key to encrypt the link.

$$SK = E_{LTK}(SKD_{master} \| SKD_{slave})$$

The initialization vector (IV) is also calculated from the two values contributed by both devices IV_{master} contributed by the master and IV_{slave} contributed by the slave by concatenating them together.

$$IV = IV_{master} \| IV_{slave}$$

All data transfers are paused while encryption is starting. This is done on the master before it sends its encryption request packet; on the slave, it's done before it sends its encryption response packet. This ensures that no data can be sent unencrypted while the encryption is starting up; it also helps the three-way handshake to perform correctly.

Once the SK and IV have been calculated by the slave, the slave sends the start encryption request packet (LL_START_ENC_REQ) unencrypted, but the slave sets to receive an encrypted packet by using the SK and IV values just calculated. If the master doesn't receive the slave's packet, the master will respond by requesting the same packet again, using an empty packet because all other data packets have been paused. Because empty packets can never be encrypted (there is no payload to encrypt), the slave can receive either this packet or the master's next encrypted packet.

The master responds to the slave's packet by sending an encrypted packet (LL_START_RSP) that uses the same SK and IV values it has just calculated and setting up to receive an encrypted packet in response. The slave can receive this encrypted packet because it was configured to receive encrypted packets. The slave will now turn on encrypted packet transmission.

Upon receipt of this master's encrypted packet, the slave will respond with an encrypted packet (LL_START_RSP). The master can receive this encrypted packet because it has already turned on the reception of encrypted packets.

Once the master has received this final packet, it can turn on the flow of application data, all of which will be encrypted. Once the slave has received the Link Layer acknowledgement of this final packet, it can turn on the flow of application data, all of which will be encrypted.

On Bluetooth classic, authentication must be performed before encryption can be started. This costs both in terms of time and additional messages that need to be transmitted. In Bluetooth low energy, this process is not necessary because each packet that is transmitted includes authentication. Therefore, the authentication of the link is done on each and every packet, not just once at the start of the encryption process.

7.10.4 Restarting Encryption

Restarting encryption is useful to refresh the session key and is used to encrypt the link when the packet counter has almost expired or because a new link key has been established, and the host wants to use this new link key to derive the session key. It is not a procedure that will be used very often; in fact, some devices might never have a connection up long enough and so never require the restarting of encryption.

To restart encryption, the same starting encryption process is used, but only after encryption is paused. Pausing encryption means that application data is paused and then encryption is turned off. This ensures that no application data can be sent unencrypted while encryption is restarting. It is effectively a two-step process: The master must first pause encryption and then restart it by using the starting encryption procedure.

The pausing of encryption is another three-way handshake, but in the opposite order to the way that the starting encryption handshake is carried out.

As illustrated in Figure 7–27, the master sends a pause encryption request packet (LL_PAUSE_ENC_REQ) to the slave, after the master's application data is paused.

Upon receipt of the pause encryption request packet from the master, the slave will pause its application data and then send a pause encryption response packet (LL_PAUSE_ENC_RSP) and disable the reception of encrypted packets; the master will only be able to send empty packets to acknowledge any packets the slave might send.

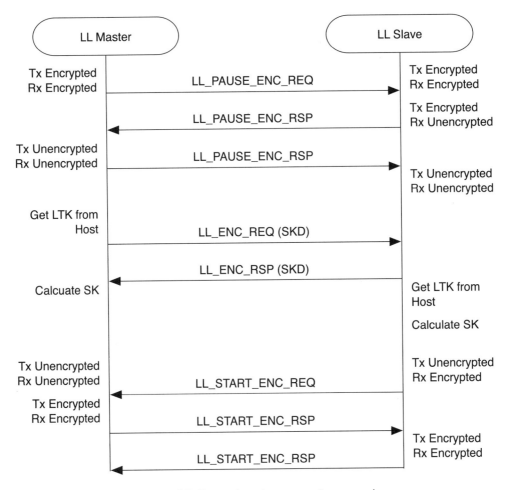

Figure 7-27 Restarting the encryption procedure

After it has received this slave pause encryption response packet, the master will turn off the transmission and reception of encrypted packets and reply with another pause encryption response packet (LL_PAUSE_ENC_RSP). The slave can receive this packet because it has already disabled encryption. Once the slave receives this packet, it will also disable encryption for transmitted packets.

Encryption is now disabled. No application data can be sent because all application data has been paused. This protects against the possibility of sensitive application data being sent unencrypted.

Once encryption has been disabled, the master will immediately send the encryption request packet (LL_ENC_REQ) to the slave to initiate the starting encryption procedure defined in the last section.

7.10.5 Version Exchange

Sometimes it is useful for debugging purposes to find out a little more about a device than just its device address and what information is available in the host and application layers. Version information is only useful for debugging purposes and can be obtained from devices by the Link Layer autonomously or by the host requesting the information, as shown in Figure 7–28. The version information cannot be used for changing the behavior of the device; therefore, it does not need to be exchanged on every new connection. However, most devices will request this information once every 10 or more connections so that a sniffer can pick this information up and help with the debugging of the connection.

This autonomous behavior of the Link Layer troubles some people. Why would a Link Layer want to receive version information every 10 connections? Simply, when you have a problem with two devices that need debugging, you might not have any ability to trigger a version information exchange from a host in either of these devices. Therefore, to be able to characterize these devices, a version information exchange that occurs infrequently is much better than nothing at all.

The classic example here is a device that is paired with only one other device (so no other device can connect and request this version information) that does not exhibit the buggy behavior for many months after being initially paired, and now that device is exhibiting buggy behavior. Having these version exchanges autonomously sent by the Link Layer every few connections will help tremendously in solving this problem now, and not in a few months' time.

Version information includes the following:

- Version number
- Company identifier
- Sub-version number

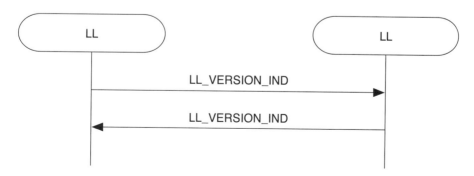

Figure 7–28 The version exchange procedure

The version number is a number assigned to a given published version of the Bluetooth specification with which this device is compliant.

The company identifier is an assigned number from the Bluetooth Special Interest Group (SIG) that is allocated to the company that manufactured this controller.

The sub-version number is a number assigned by the company that manufactured the controller; it must be different for each implementation or revision of the controller. There is no defined way to assign this number, and if debugging of a device is important, then the manufacturer of the device should be contacted directly. The device manufacturer contact details are typically exchanged at UnPlugFest events, where engineers test and validate these controllers before shipping them in mass volume. This contact information is also available in the qualification database for all shipping products.

To exchange version information, either device can send a version information indication packet, if they have not already done so during the current active connection. If a device receives a version information indication from a peer device and it has not already sent its own message, it will respond with a version information indication packet. After this has completed once, the procedure cannot be requested again. This information is static and therefore will not change while in a connection. So, it's pointless to ask again for the same static information. The controller should cache this information while the connection is active in case the host asks for it again.

7.10.6 Feature Exchange

The feature information is used by a peer device to determine what a device can do. It exposes the set of optional features supported by a device. As illustrated in Figure 7–29, this information can be requested by the master by using the feature request packet (LL_FEATURE_REQ) at most once, if at all, during any connection. Upon receipt of the master's request, the slave replies with the feature response packet (LL_FEATURE_RSP). This information does not need to be exchanged to attempt to use an optional feature because all features must be enabled, and all optional features have a way to reject this request. For example, if encryption is not supported or enabled, then a reject indication packet is returned, safely rejecting the encryption request.

In the first version of Bluetooth low energy, there is only one feature bit defined: encryption supported. This states that encryption is supported by a device.

7.10.7 Terminating Connections

The only other Link Layer control procedure that can be performed is the termination procedure. This, unsurprisingly, disconnects the link and moves both the master and

7.10 Managing Connections

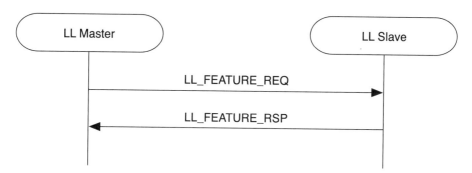

Figure 7-29 The feature exchange procedure

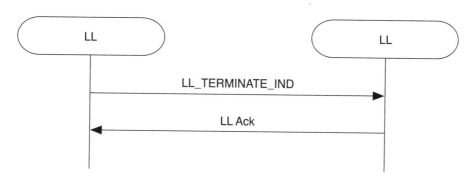

Figure 7-30 The terminate procedure

slave connections back into standby state. A link can be terminated at any time, by either device, for any reason, as shown in Figure 7-30.

To terminate the link, a device sends a terminate indication packet (LL_TERMINATE_IND). It then waits for the acknowledgement of this packet at the Link Layer, after which it disconnects. If a device sends the terminate indication packet but does not receive any acknowledgement of this packet, the sending device will just timeout this procedure and the connection will still be disconnected. This timeout is the same as the supervision timeout value.

As soon as a device receives the terminate indication packet, in response it sends an empty packet to immediately acknowledge this packet, and then it disconnects.

Connections can also be terminated for two other reasons:

- Supervision timeout
- MIC Failure

When these other reasons are encountered, there is no terminate indication packet sent; the link is just disconnected, and the hosts on both sides are notified accordingly.

7.11 Robustness

To be robust, the Link Layer uses two very strong algorithms to ensure that data gets through without interference and that when data is sent from one host, it is delivered to a peer host unchanged.

7.11.1 Adaptive Frequency Hopping

Adaptive frequency hopping is essential for the efficient movement of data between devices communicating by using the 2.4GHz ISM band. There are so many different devices using this band that having technology that protects against interference is essential. Adaptive frequency hopping was originally brought into the Bluetooth classic specification in late 2003; it brought in a step change in performance improvements when compared with pre-2003 devices.

Simply put, adaptive frequency hopping is a way of taking a frequency-hopping radio and masking out bad channels by remapping them onto good channels. To do this, both devices have a channel map. This channel map is 37 bits in length, with each bit mapping directly onto a given Link Layer channel. Bit 0 in the channel map is related to the Link Layer channel 0; bit 1 in the channel map is for Link Layer channel 1; bit 36 in the channel map is related to the Link Layer channel 36. If a channel's bit is set to one, the channel is a used channel; if a channel's bit is set to zero, the channel is an unused channel. A minimum of 2 bits must be set as used into this channel map.

Take, for example, the channel map if all three main Wi-Fi channels are actively being used. These are channels centered at 2412MHz, 2437MHz, and 2462MHz. This means that only 9 Link Layer channels would be considered good: 9, 10, 21, 22, 23, 33, 34, 35, and 36. Three values define the channel map, *ChannelMap* is transferred between devices, *Used* is the set of good channels from the *ChannelMap*, and *numUsed* is the number of these good channels. These values for the preceding example would therefore be as follows:

$$ChannelMap = 0001111000000000111000000000011000000000_2$$

$$Used = [9, 10, 21, 22, 23, 33, 34, 35, 36]$$

$$numUsed = 9$$

If a connection were set up with this channel map, sent in the connect request packet from the master to the slave, with the hop interval set to 7, the preadaptive frequency-hopping channels used for this connection would be calculated by using the frequency-hopping equation, as shown in Figure 7–5, and as explained

7.11 Robustness

in Section 7.4.1.

$$f_{n+1} = (f_n + hop) \bmod 37 \qquad (7\text{-}5)$$

The adaptation of these channels occurs after the initial frequency f_n is calculated.

The first connection event on the link should be channel 7. This is not in the set of good channels, so it would be remapped into the set of good channels. To do this, the set of good channels, *Used*, is ordered by channel number, and the unmapped channel number is mapped into this by modulating this with the number of good channels, *numUsed*: in this case, we have 9 good channels. The seventh good channel, Link Layer channel 35, is used as the channel in the connection event.

$$Used[7 \bmod numUsed] \Rightarrow 35$$

At the next connection event, the channel would be calculated from the last unmapped channel.

$$f_{n+1} = (f_n + hop) \bmod 37$$
$$f_{n+1} = (7 + 7) \bmod 37$$
$$(7 + 7) \bmod 37 = 14 \bmod 37 \Rightarrow 14$$

Channel 14 is also an unused channel, so this is again remapped into the *Used* channels.

$$Used[14 \bmod numUsed] = Used[5] \Rightarrow 33$$

At the next connection event, the channel would be calculated again from the last unmapped channel:

$$(14 + 7) \bmod 37 = 21 \bmod 37 \Rightarrow 21$$

Channel 21 is in the set of used channels, so this is just used directly.

As you can see in Table 7–5, this can continue for each connection event in the future.

By using adaptive frequency hopping, data can always be transferred, even if lots of interferers are in the area, by remapping channels that are marked as unused to used channels that are known to be good.

Table 7–5 Remapping Channels by Using Adaptive Frequency Hopping

Connection Event Counter	Unmapped Channel	Used/Unused	Remapped Channel
0	7	Unused	35
1	14	Unused	33
2	21	Used	21
3	28	Unused	10
4	35	Used	35
5	5	Unused	33
6	12	Unused	22
7	19	Unused	10
8	26	Unused	36
9	33	Used	33
10	3	Unused	22
11	10	Used	10

7.11.2 Strong CRCs

The size of the CRC value in Bluetooth low energy is 50 percent larger than the size of the CRC in Bluetooth classic and most other wireless short-range devices. Unfortunately, Bluetooth classic has proven that a 16-bit CRC is not strong enough for the 2.4GHz ISM band. This was solved by adding another 16-bit CRC in the Logical Link Control and Adaptation Protocol Layer (L2CAP) to ensure that even if the CRC protecting data in the controller passed a packet up to L2CAP that has bit errors that were not protected, the host can still protect against these bit errors by using this second CRC. Unfortunately, this makes every L2CAP packet much larger, and for resource-constrained devices, the complexities of calculating a CRC in software also complicate the implementation.

Because of the experience gained from Bluetooth classic, Bluetooth low energy uses a much stronger CRC at the controller. This CRC is a 24-bit value that protects the packet. This CRC will detect all single-bit, two-bit, three-bit, four-bit, and five-bit errors, and all other odd numbers of bit errors in the packet. This is much stronger than a 16-bit CRC, and when compared with much larger packets used in wired protocols such as Internet Protocol (IP), it provides the same strength. This is because the packets in low energy are much shorter; thus, a shorter CRC can provide an equivalent level of protection.

The other problem that Bluetooth classic has is that the header of its packets is protected by a relatively weak header error-check value, which is just an 8-bit CRC. Obviously, if this falsely identified a packet, the main CRC should also fail.

Again, this experience has shown that having a separate header error check and main packet CRC does not help much. If the packet is good, then the controller still has to run two separate LFSRs, one for the header and one for the payload. If the packet is bad, the probability of the error occurring in the header is relatively low; the whole packet will probably have to be received anyway, which will not result in the expected power savings by using a separate header error check in the first place.

Bluetooth low energy takes an alternate design choice. It uses a single strong CRC that protects against the header, length, and payload fields of the packet, including any encrypted packets' message integrity check value. This means that the full strength of the strong CRC is used over the complete packet, except for the preamble and access address; both of which must be received bit for bit perfect for the packet to be received in the first place. This is not only more robust but equally efficient.

And by having just one CRC—as compared to the three used in Bluetooth classic—the system is also simpler to implement. This shows why designing all layers in the specification at the same time is the best way to create a top-to-bottom, high-quality specification that can meet the market requirements.

7.12 Optimizations for Low Power

The following sections discuss how Bluetooth low energy has been optimized for ultra-low power consumption. The primary methods include the following:

- Keeping the packets short
- Using a high physical bit rate
- Providing low overhead
- Optimized acknowledgement scheme
- Single-channel connection events
- Subrating connection events
- Using offline encryption

Power consumption can be measured in multiple ways, but here it will be considered that two types of power consumption are critical to low power consumption within a device. First, low peak-power consumption is essential to allow a device to be powered from low-cost button-cell batteries. If the peak power is too high, the batteries would burn out too quickly, significantly reducing the lifetime of the device. Second, low power-per-application-bit is essential to allow a device to be used for a long time, sending a certain quantity of application data.

7.12.1 Short Packets

One of the exceptionally complicated parts of wireless technology is the actual radio that is used. Most of these radios are built by using bulk CMOS[2] technology. This creates a dilemma for designers because to make the radio stable, they need to increase the cost by adding circuitry to keep the frequency stable. Bluetooth low energy solves this for them because the packet length is sufficiently small that this heating effect is minimized. It does not need a very long packet to cause this problem. The 3-millisecond packets in Bluetooth classic are long enough to cause problems.

This very simple design decision emphasizes the level of detail that the designers of Bluetooth low energy have taken, optimizing the Link Layer specification by taking into account the physical properties of the silicon manufacturing processes used.

If the packets are never more than a few hundred microseconds in length, then no calibration of the radio or stabilization circuitry will be required. The frequency can drift for this period of time without concerns that it will drift outside the frequency drift requirements stated in the specifications. In Bluetooth low energy, the longest possible packet is 376 microseconds; this is short enough that the heating up of the silicon will not change the frequency of transmitted packets enough to drift outside the limits allowed. While in a connection, the longest possible packet is smaller at just 328 microseconds, as depicted in Figure 7-31.

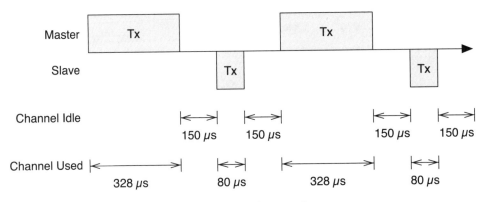

Figure 7-31 Short packets

---

2. Complimentary Metal on Silicon (CMOS) is used to manufacture over 95 percent of all silicon chips. It is a very low-cost technology.

Therefore, by keeping packets short, there is no need for constant calibration of the radios. This reduces peak power consumption by reducing the quantity and complexity of circuitry that is required to be powered during a packet transmission or reception.

It should also be noted that after transmitting a very long packet, a gap of 150 microseconds is required. This interpacket gap allows the silicon to cool down between packets. Thus, allowing no calibration of frequencies needed between transmitting and receiving or receiving and transmitting packets, further reducing power consumption. This means that when transmitting data in one direction on an encrypted link, the maximum duty cycle is just:

$$\frac{maximum\ size\ packet + acknowledge\ packet}{total\ time\ to\ send\ and\ acknowledge\ data}$$

$$\frac{328 + 80}{(328 + 150 + 80 + 150)} = \frac{408}{708} = \sim 58\%$$

A 58 percent duty cycle is very low for a wireless technology. Bluetooth classic has a duty cycle of 72 percent, whereas very high-speed wireless technologies will have duty cycles in the high 90 percent range. Bluetooth low energy is optimized for small discrete pieces of data being sent, not for the highest possible throughput of data.

7.12.2 High Bit Rate

When transmitting data, a radio requires a large amount of current. Most of this current is consumed by the requirement to run a 2.4GHz oscillator on which the radio signal is modulated. Running very high frequencies in CMOS requires very high currents. CMOS is optimized for gates that don't change their state all the time because most gates in a digital system do not change state all the time. Any radio running at 2.4GHz built using a bulk CMOS process will therefore use similar current generating this 2.4GHz signal.

Given that it is almost impossible to reduce the power consumption below a certain level because of the oscillator, the efficiency of the modulated signal becomes significant. The quicker you can transmit a given amount of data, the more efficient your radio will be. Bluetooth low energy transmits data at 1,000,000 bits per second.

For example, if a device using another technology can only transmit at a quarter of the rate of Bluetooth low energy, it would take four times as long to transmit a given sequence of bits. Thus, it would take four times as much power to transmit the same amount of data as a Bluetooth low energy system would take. Consequently, high bit rates are a good thing.

It should also be considered that it is possible to go too far on the data rate front. A complex modulation scheme that can transmit, for example, 10 times as many bits per second as Bluetooth low energy would typically take much more power to modulate and demodulate these bits. So, although these bits might take less time to transmit, the extra energy used for modulation means that the power per bit is about the same. However, the current used when receiving nothing, due to clock drift, is significantly higher. Therefore, a simple modulation scheme with low peak-power consumption with high data rates is the most efficient way of sending data. This is the sweet spot that the designers of Bluetooth low energy used.

7.12.3 Low Overhead

Given that every bit counts, the amount of overhead is critically important when considering a radio technology. In Bluetooth low energy, the overhead includes everything that cannot contain application data. This includes the preamble, access address, header, length, the CRC fields, and the optional MIC value.

For an unencrypted packet, the efficiency measured as the size of application data compared with the total packet size required to transmit this data goes from 29 percent to 73 percent, as demonstrated in Table 7–6. The larger the packet, the more efficient the radio is in terms of overhead. For encrypted packets, the efficiency is lower, primarily because of the extra 4 octets of MIC included in each packet. This efficiency is very good, compared with the efficiency of similar proprietary radios. For example, Zigbee has a packet overhead of anywhere from 15 to 31 octets before any encryption overhead. When considering the four-times slower physical bit rate for Zigbee, a short four-octet application data in Zigbee could require up to 10 times more energy than a Bluetooth low energy solution.

Table 7–6 The Overhead for Application Data

Packet Type	Application Data Size (octets)	Overhead (octets)	Efficiency (%)
Unencrypted	4	10	29%
Unencrypted	8	10	44%
Unencrypted	16	10	62%
Unencrypted	27	10	73%
Encrypted	4	14	22%
Encrypted	8	14	36%
Encrypted	16	14	53%
Encrypted	27	14	66%

7.12.4 Acknowledgement Scheme

One interesting side effect of the Link Layer acknowledgement scheme is that it does not require that the acknowledgement of a packet be performed, or even delivered, immediately. This is a radical difference between Bluetooth classic and Bluetooth low energy.

In Bluetooth classic, the receiver must acknowledge the packet at the next opportunity it has to transmit. If the acknowledgement is not received immediately, the receiver must signal a negative acknowledgement in the next packet it transmits. This causes the most problems when synchronous links are active and the slave is attempting to send data.

In Bluetooth low energy, every packet sent can acknowledge the last packet transmitted, even if this was transmitted some time ago. This means that devices never have to transmit immediately to send their acknowledgements. The device can choose to wait until it has data to send (or needs to transmit for some other reason, such as timeouts) before acknowledging the last packet. This allows for very fast and efficient acknowledgements that are required when transmitting large quantities of data very quickly.

7.12.5 Single-Channel Connection Events

All communication between a master and a slave occurs in a connection event. A connection event is a packet transmitted by the master, followed by a series of alternating packets sent by the slave and master. In Bluetooth v1.1, every single packet transmitted either as master or slave was transmitted on a different channel. If the master to slave packet was transmitted on one channel, the response would have been sent on a different channel. The problem with this is that even though the slave might have received the master's transmission perfectly, the next channel that was used for the reply might have interference and, therefore, the acknowledgement for the master's data packet might not get through.

When it added Adaptive Frequency Hopping, Bluetooth v1.2 made each slave transmission use the same channel as the master's transmission. In the preceding example, the slave would have replied on the same channel that the master used to transmit its data, and therefore that packet would most likely get through; the assumption is that if a packet is good when data is transmitted from the master to the slave, it will be good when data is transmitted from the slave to the master. This is a good assumption and has improved the actual data rates in heavy interference situations.

Bluetooth low energy has taken this to the logical conclusion. If a channel is good, then why stop using it? If a channel is good, then it should be used for as long as

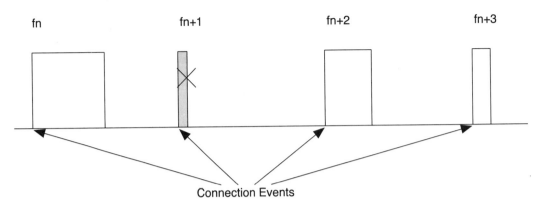

Figure 7-32 Single-channel connection events

possible. The ultimate invalid conclusion would be to just stop frequency hopping altogether. After all, if the data gets through, then you should just stay on that frequency until it stops working, and then change to another channel at that point.

Unfortunately, at the point when you do need to change frequency, you cannot send any signaling packets to coordinate that change. It also uses a lot of power resynchronizing the single frequency once the channel starts to fail. Given the transient nature of interferers in a location, especially with the bursty nature of much Internet traffic over a Wi-Fi network, this model breaks down very quickly.

This one-channel model also reduces the number of co-located networks because each will naturally drift to a clean frequency. If the number of channels that are clean is small, the number of co-located networks will also be small. A frequency-hopping algorithm distributes network traffic in both time and frequency, allowing for many more simultaneous networks to be active in the same area at the same time. The alternate algorithm—stay on one frequency until it breaks—doesn't work.

The low-energy approach is to stay on one frequency in a single-connection event, as if you know it works and can send data and acknowledgements, and then move to a new frequency at the next connection event. This means that at any point in time, the frequency to use is fully deterministic. If a particular channel does not work, the two devices will stop using that channel very quickly and resynchronize at the next connection event on a different channel and continue sending data (see Figure 7-32). This means that data can flow with the minimum of latency impacts even where there is lots of interference that is highly unpredictable—a situation that is common in many homes and businesses.

7.12.6 Subrating Connection Events

In many situations, the master device has many more resources than the slave device. A computer keyboard has a much smaller battery than the computer, and although

7.12 Optimizations for Low Power

the computer might be recharged regularly, the battery in the keyboard is expected to last for years. However, many devices have very small latency requirements. When a user presses a key on a keyboard, that keystroke needs to be delivered to the peer device as quickly as possible. Low latency is a requirement for user interaction. However, the conflicts of low latency and low power have to be resolved.

Low power requires the slave device to only listen for the master device very infrequently. Low latency requires the slave device to be able to transmit as often as possible. Given that the slave cannot talk unless the master speaks to it in a connection event, low latency requires the master to continuously poll the slave to see if it has any data to send. However, if the slave had to listen to each of these connection events, it would not be low power.

The solution is to allow the slave to ignore most connection events from the master, as illustrated in Figure 7–33. The slave would still be able to synchronize and send data at the earliest possible opportunity once it needs to send some data, keeping the latency of data to the connection event interval to that of the master. However, the slave would also be able to ignore a certain number of connection events; this is called the *slave latency*. The number of connection events that can be missed determines the amount of power that the slave can save. The more connection events that can be missed, the lower power the slave can be.

There is a limit to the slave latency. It is not possible to have a slave latency that is longer than the supervision timeout of the connection. It is also not recommended to have a slave latency that gives fewer than 6 opportunities for the slave to resynchronize to the master. For example, if the supervision timeout is 5 seconds, and the connection event interval was 50 milliseconds, then a slave latency of 4,650 milliseconds would give the slave 6 opportunities to resynchronize at the end of the

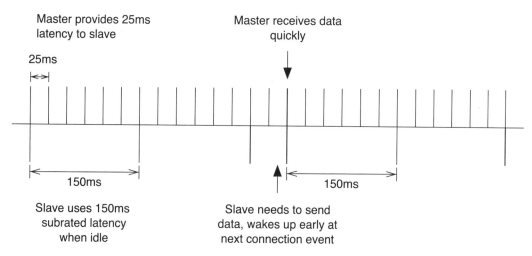

Figure 7–33 Subrated connection events

slave latency. It would also allow the slave to communicate just once every 4.5 seconds, a duty cycle of just 0.0069 percent but still allowing the slave to transmit data to the master in an average time of just 25 milliseconds.

7.12.7 Offline Encryption

Encrypting data is typically considered a very high-power procedure. A device must compute cipher blocks for the payload and message authentication codes as well as compute the authentication code itself. This requires, for a maximum-sized packet, a total of seven iterations of the AES-128 encryption block. If this were required to be performed in real time, the peak current of the packet transmitter and receiver would be significantly higher.

In Bluetooth low energy, the encryption of data and authentication code can be computed in the background. Before a packet is transmitted, the encryption of the data can be performed when the radio is still off. This encryption of data doesn't depend on the sequence number, the next expected sequence number, or the more data bit. Therefore, the data can be encrypted at any time, from the point it reaches the Link Layer, to just before being transmitted. Also, this data can be retransmitted any number of times, and the encryption and authentication code will not change for each retransmission, even if the next expected sequence number or more data bit changes. This lowers the peak power consumption, and removes the cost of retransmitting encrypted packets.

When receiving encrypted data, the CRC value is computed in real time and is the only value that determines whether the data was received correctly. The encrypted data can be kept in the Link Layer until radio activity stops and there is spare power to decrypt the packet. This decryption of the packet can be done at any time before the packet is delivered to the host. This reduces the peak power for a receiver. Also, if the packet is retransmitted, then upon reception because the sequence number is repeated, it is not necessary to decrypt the packet again. This lowers peak power consumption and removes the cost of re-receiving encrypted packets.

Chapter 8
The Host/Controller Interface

The best way to predict the future is to invent it.
—*Alan Kay*

8.1 Introduction

As shown in Figure 3–1, the Host Controller Interface (HCI) is the interface between the host and the controller. It is responsible for two main functions: It sends commands to the controller and receives events back, and it sends and receives data from a peer device.

Typically, the host interface is both a physical and logical interface between devices. The logical interface defines a number of packet formats for commands, events, and data. The physical interface then defines how these packets can be transported between the host and controller.

8.2 Physical Interfaces

There are four defined physical interfaces in the Bluetooth specification. Each is optimized for a different purpose:

- The Universal Asynchronous Receiver/Transmitter (UART) is optimized for very simple implementations.

- 3-Wire UART is optimized for reliable UART implementations.

- USB is optimized for high speed, typically in computers and similar devices.

- Secure Digital Input Output (SDIO) is optimized for medium speed, typically in consumer electronic devices.

8.2.1 UART

The UART interface is the simplest of all the transports available. It defines a simple universal asynchronous receiver/transmitter that is connected in a null-modem configuration between the host and the controller. Null modem simply means the transmit and receive wires cross over at opposite ends of the link so that transmit data (TXD) from the controller connects to receive data (RXD) on the host, and vice versa. Three-wire UARTS only have TXD, RXD, and ground. Five-Wire UARTS also have flow control wires by which the request to send (RTS) on the controller connects to clear to send (CTS) on the host, and vice versa.

In Bluetooth low energy, the UART interface always uses 8-bit characters with no parity and one stop bit. The stop bit is an extra bit's worth of time at the end of each byte. The parity bit is used for error checking when used. When the link between Bluetooth controller and host is very short, errors don't tend to occur and the parity check is not needed. It uses hardware flow control lines CTS and RTS. To send an HCI packet, one of the following three packet type codes is prepended to it. That is it. That's all it is.

- Command = 0x01
- Data = 0x02
- Event = 0x04

Unfortunately, the UART interface cannot do any low-power signaling, and, therefore, this interface is not suitable for very-low-power devices. Some devices will use additional hardware signaling lines to allow the interface to be moved into a very-low-power mode, but these are typically proprietary extensions and each device will do these extensions differently.

In Bluetooth classic, the HCI packet type of 0x03 is also defined for synchronous data packets; low energy does not use synchronous data packets; thus, they would never be sent to or from a low-energy-only controller.

8.2.2 3-Wire UART

The 3-wire UART is a little more complex than the previously described UART transport because it is designed to work without any hardware flow control lines as well as in the presence of some bit errors. If your host and controller are separated by more than a few millimeters and they are in a noisy electrical environment, using the 3-wire UART is a strong recommendation.

For the 3-wire UART to work, channels are used in a similar manner to the UART interface described earlier. The channel numbers are the same, with 0x1 for

8.2 Physical Interfaces

commands, 0x2 for data, and 0x4 for events. The 3-wire UART also defines two additional channels for link establishment on channel 0xF and one for acknowledgements on channel 0x0. Note: Channel numbers in 3-wire UART are only four bits in size.

The 3-Wire UART has three main modes:

- Link establishment
- Active state
- Low-power state

The link establishment channel is used to confirm that the peer device is awake and to configure any parameters. The link establishment channel can also be used for automatic baud-rate detection. It does this by sending link establishment messages at different baud rates and determining which one caused the peer device to respond. The link establishment process is effectively a three-way handshake between the two devices to move them into the normal state. This includes configuration of the reliable sliding window size, whether a cyclic redundancy check (CRC) is used, and if out-of-frame software flow control is used.

In the active state, reliable packets can be sent. All packets are framed and include a sequence number and an acknowledgement number. These numbers are both three bits in size, allowing multiple packets to be in transit at the same time. This is useful for very fast UARTs for which it might take some time to process a packet.

The framing of packets uses SLIP, as defined in RFC 1055. SLIP places a 0xC0 octet at the start and end of every packet to frame it. It then replaces any occurrences of 0xC0 in the packet with a two-octet byte stream of 0xDB 0xDC. Given that 0xDB is an escape sequence, this also needs to be converted into a two-octet byte stream of 0xDB 0xDD. If out-of-frame software flow control is used, the XON and XOFF octets will also be escaped to 0xDB 0xDE and 0xDB 0xDF.

The packet header of a framed packet also includes the length of the packet and a header checksum. The header checksum just validates that the header information can be trusted. If it is invalid, then the entire packet is rejected, and a retransmission scheme will automatically retransmit this packet.

The payload can be up to 4,095 bytes and is protected by a CRC value. The CRC used in the 3-wire UART is the same CRC that is used in the Bluetooth classic baseband packets—the 16-bit CRC-CCITT. Again, if the CRC fails to validate the packet, the packet is just ignored, and the entire packet will be retransmitted by the peer device.

It is also possible to move the connection into low power mode by sending a "sleep" message. In this mode, the UART is typically turned off and any packet that is transmitted is not guaranteed to be fully received; it can take some time for the

UART hardware on the peer device to wake up once a packet is transmitted. To aid the efficient wakeup of a peer device before sending HCI messages, a very short wakeup message can be sent. The device responds with a "woken" message, after which, the devices have an active connection and can send any packets.

If a device only has a UART interface, and the basic UART interface isn't proving robust against bit errors, the 3-wire UART is the right way to go.

8.2.3 USB

The USB interface is optimized for devices that already include a USB host. It defines how to transmit commands and events as well as data between the host and the device. For enumeration, a standard class code for Bluetooth devices is also defined. This has allowed plug-and-play Bluetooth dongles to be sold by many manufacturers.

Commands from the host to a device are sent on the Control endpoint (0x00) using a "host to device class request, device as target" request type. Events are polled from the device by the host using an Interrupt endpoint (0x81). This endpoint should be polled every millisecond to ensure that events can flow with the minimum of latency.

Data is sent on two endpoints: one for Bulk data out (0x02) for data from the host to the device, and one for Bulk data in (0x82) for data from the device to the host. Again, a one-millisecond poll interval is used; however, multiple Bulk data USB packets can be sent in a single frame, allowing for very high data throughput where needed.

The biggest problem with USB is that it is not very power sensitive. To implement a USB interface requires lots of high-speed hardware and software control. This is an expensive proposition. Another problem is that the USB host must poll for data from the device every millisecond. Typically, this stops the host from moving into the lowest-power processor states while the device is being used. This can be solved by another USB featured called *Link Power Management*. It is recommended to use this on Bluetooth devices.

8.2.4 SDIO

The SDIO protocol is a high-speed interface by which a host can communicate with a controller over the SDIO Card Type-A interface. SDIO is a packet-based bus that uses between 4 and 8 lines to transmit data in both directions very quickly using very little power.

The transport uses the same channel assignments that are used by UART and 3-wire UART transports for commands, events, and data.

The SDIO interface has very low error rates, which makes it very useful for devices that already have an SDIO interface available.

To obtain the full SDIO transport specifications, you must be a member of the SD Association. However, a version of these specifications has been made available to help companies evaluate the technology. These simplified specifications do not include everything that is necessary to build a product, but they do contain enough information to obtain an understanding of how the system works.

8.3 Logical Interface

Above the HCI physical transports is the logical interface. It is a logical interface because in the single-chip device, this interface doesn't need to be implemented as a message-passing interface between components. For a system in which the controller and host are on separate chips connected by a physical interface, the HCI logical interface is represented as physical packets transferred over this physical interface.

There are three concepts to understand about the logical interface:

- Channels
- Packet formats
- Flow control

8.3.1 HCI Channels

Whenever a controller has a connection to another device, the controller's lower HCI interface creates an HCI channel that is identified by using a connection handle. This connection handle is used to identify all data that is sent from the host to the controller that is to be sent to a specific peer device as well as all data that is received from that peer device by the controller before it is sent to the host.

The connection handle is given to the host whenever the attempt to create a connection completes. The connection handle is valid until the connection is terminated, either locally, by the Link Layer termination procedure, or due to a link supervision timeout.

8.3.2 Command Packets

To command the controller to do something useful, command packets are sent by the host to the controller. These command packets typically are used to either configure the state of the controller or ask it to do something.

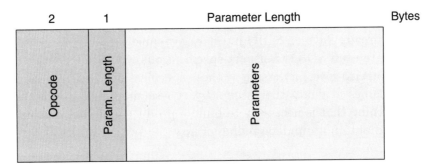

Figure 8-1 The HCI command packet format

As illustrated in Figure 8-1, the HCI command packet contains an opcode that determines the command that is being sent, a parameter length field, and the parameters for the command. Each command has its own unique set of parameters.

There are three basic types of commands in Bluetooth low energy that perform the following functions:

- Configure the controller state
- Request a specific action
- Control a connection

8.3.2.1 Configuring the Controller State

The controller can be considered to be one big state machine that has a number of parameters which can be configured. For example, advertising can be considered to be a state that has the following state that can be configured by using the LE Set Advertising Parameters command, LE Set Advertising Data command, LE Set Scan Response Data command, and the LE Set Advertise Enable command.

Typically, within Bluetooth low energy, the state used within a state machine cannot be changed while that state is being used. Therefore, it is impossible to change the advertising parameters while advertising is enabled. It is therefore necessary to disable advertising, change the advertising parameters, and then enable advertising again.

8.3.2.2 Requesting a Specific Action

Some commands request a specific action to occur without altering the state of the device or the state of a connection. For example, the LE Encrypt command takes a key and some plain-text data and requests the controller to generate some encrypted data based on these.

8.3.2.3 Controlling a Connection

When a connection has been created between two devices, commands can be sent to manage this connection. These commands always include the connection handle. For example, the LE Read Channel Map command is used to read the current adaptive frequency-hopping channel map for a given connection.

8.3.3 Event Packets

Event packets are used to send information from the controller to the host, typically in response to something that the host has previously commanded.

Figure 8–2 shows that the HCI event packet contains an event code that determines which event is being sent, a parameter length field, and the parameters for the event. Each event has its own unique set of parameters.

Three basic event types are used in Bluetooth low energy:

- Generic command complete events
- Generic command status events
- Command-specific completion events

8.3.3.1 Generic Command Complete Events

Whenever a command is sent to the controller that can be completed immediately, a generic command complete event is returned. This is the Command Complete event. This event is generic in that the event parameters include the command opcode that this event is completing along with the return parameters that are specified by that command. The first parameter of all command return parameters is a status code that specifies whether the command was successful or had an error and couldn't complete.

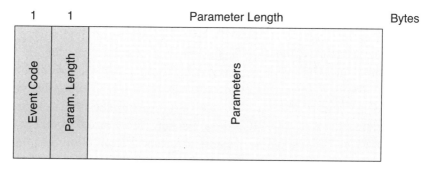

Figure 8–2 The HCI event packet format

For example, the LE Rand command that is used to command the controller to return a random number has two return parameters: the status code and the requested random number.

The generic command complete event is used whenever the controller can complete a command without doing any over-the-air transactions. For example, the LE Encrypt command does not request any Link Layer packets to be transmitted, so the generic command complete event is used, whereas the LE Create Connection command requires that at a minimum a Link Layer CONNECT_REQ packet is transmitted before the connection is established; therefore, the generic command complete event cannot be used.

8.3.3.2 Generic Command Status Events

For those commands that perform over-the-air transactions, such as creating the connection mentioned in the previous section, a generic command status event is normally followed some time later by a command-specific completion event. The generic command status event is the Command Status event.

8.3.3.3 Command-Specific Completion Events

Some commands require time to complete. These commands always have command-specific completion events. For all of these commands, there is just one command completion event. For example, the LE Create Connection command will first have a Command Status event sent and then an LE Connection Complete event will be sent once the connection is created or the connection fails. It is not until this command-specific completion event is received that the command is considered to be finished.

8.3.4 Data Packets

Data packets are used to send application data from the host to the controller so that it can be transmitted to a peer device, and from the controller to the host that has been received from a peer device.

As illustrated in Figure 8–3, data packets are always labelled with a connection handle. This is a 12-bit value that is provided to the host in the LE Connection Complete event. Until the event is received by the host, it cannot send any data to a peer device. However, once this event is received, the host can start to send data to the peer device, and it will start to receive data from the peer device.

There are two flags in the HCI data packet: the Packet Boundary Flag and the Broadcast Flag. Because these packets are reused from Bluetooth classic, some of these flags don't have any meaning in low energy. The Packet Boundary Flag determines if this packet is a start or continuation of a higher-layer (Logical Link

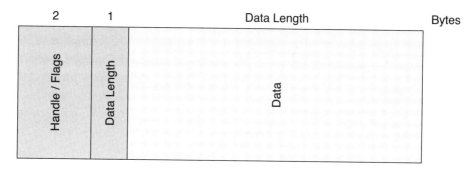

Figure 8–3 The HCI data packet format

Control and Adaptation Protocol) message. This can be considered to be very similar to the LLID bits in a Data Channel PDU in the Link Layer (for more information, see Chapter 7, The Link Layer, Section 7.8.2).

The interesting thing is that from the host to the controller, this can contain the values 00 (start) and 01 (continuation), whereas from the controller to the host, this can contain the values 10 (start) and 01 (continuation). This is because in Bluetooth classic the value 10 from the host to the controller means that the data packet can be flushed if necessary; with Bluetooth low energy, the concept of flushing doesn't exist. Instead, it simply drops the link if the data cannot get through, so this value cannot be used. It is for this reason that the value used for the start indication from the host to the controller is 00 (instead of the LLID value 10 at the Link Layer) because this is used for unflushable data in Bluetooth classic.

8.3.5 Command Flow Control

The HCI interface has two forms of flow control: command flow control and data flow control. Command flow control is used to enable the controller to manage how many HCI commands it can process at the same time. The easiest way to consider how this works is to think of the controller as having enough memory to buffer a small number of commands. It communicates the number of these buffers to the host so that the host knows how many commands it can send to the controller at the same time.

There is no event flow control. It is assumed that the number of events that can be sent is limited by the number of commands that can be processed. It is also assumed that the host has more resources than the controller; therefore, it can buffer and process these events in sequence.

To enable command flow control, all Command Complete and Command Status events include a parameter called Num HCI Command Packets. This parameter indicates how many command packets can be buffered in the controller. Each time a command is sent, it uses one of these slots. Each time a Command Complete or Command Status event is sent to the host, it includes the number of slots that are currently free. It is also possible to send a Command Complete event for the opcode "No Operation" with a new Num HCI Command Packets at any time. This is most useful at the initial boot-up of the controller, when it might want to grant the controller more command buffer slots to speed up the initial configuration of the controller by the host.

8.3.6 Data Flow Control

Data flow control is performed in a similar manner. There are two flows of data: host to controller and controller to host. Host to controller data flow control must be used, but the flow control from the controller to the host can be ignored. Most hosts should be able to cope with the quantity of data being sent from the controller to the host, so flow control is not necessary.

For host to controller data flow control, the controller is considered to have a number of buffers, each a fixed size. Each time a data packet is sent from the host to the controller, it uses one of these buffers. Each time a data packet is successfully transmitted by the controller to the peer device, the buffer that held that data packet is released back to the host to fill up with another data packet.

A slight complication with this flow control system is that a dual-mode controller can have two different buffers, one for basic rate data and one for low energy data. This means that there are two HCI commands to discover the buffers on a controller: Read Buffer Size command and LE Read Buffer Size command. Using these two commands, the host can therefore determine which buffers are available. For a low energy–only device, only the LE Read Buffer Size command will return non-zero length buffers.

After the host has discovered the number of buffers available, each time it sends a data packet to the controller, it uses one of these buffers. The buffers are released when a Number Of Completed Packets event is sent from the controller to the host. This has a list of connection handles and the number of packets that was sent. This means that not only does the host know how many buffers have been released but also which data was sent to the peer devices.

8.4 Controller Setup

There are a number of things that the host can do before it attempts to transmit or receive any packets from peer devices. This includes resetting the controller to a known state, reading the device address, setting event masks, reading the flow control buffers, reading the supported features for the local controller, generating a

8.4 Controller Setup

random number, encrypting some data, setting the random address, and configuring white lists.

8.4.1 Reset the Controller to a Known State

It is always useful to reset the controller to a known state before doing anything else (see Figure 8–4). The controller might have been doing something else, and the host transport might only now have been connected. Therefore, resetting should place the controller to the standby state and return all configurable parameters to their default state.

To reset the controller, the host sends the Reset command to it. Once the controller has been reset, the Command Complete event for the reset command is sent to the host. It should be noted that this command does not reset the physical transport. If the physical transport needs to be reset, this should be done by using that transport's reset procedure. It should be further noted that even if the host can send multiple commands, due to command flow control, it is not allowed to send any other commands while the controller is resetting. Any commands that were being processed when the reset is sent will be discarded.

8.4.2 Reading the Device Address

Many low energy devices will have a preprogrammed device address. It is useful for the host to be able to read this address.

To read the device address, the host sends a Read BD_ADDR command to the controller, as shown in Figure 8–5. This causes a Command Complete event to

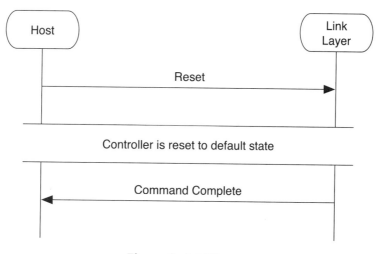

Figure 8–4 HCI reset

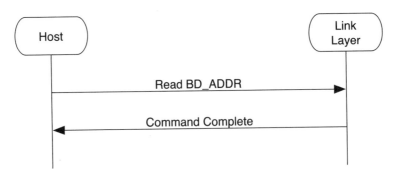

Figure 8–5 Reading the device address

be sent back with the fixed device address. If the controller does not have a fixed address, this address will have the value 00:00:00:00:00:00. The host will then need to generate and program a random address into the controller before it does anything that would cause packets to be transmitted.

8.4.3 Set Event Masks

There are many possible events defined for the controller that can be sent. In the highly likely event that additional functionality is defined in a future version of the specification, there must be a way for the controller to know which events that host is able to receive and process. If the controller just sent every event it knows to the host, but the host doesn't understand these events, then interoperability problems will occur. The only way to solve this is to allow the host to configure the controller with the set of events that it can accept. The controller will then send only those events.

As demonstrated in Figure 8–6, to set the event masks, two commands are required. First, there is the classic Set Event Mask command that configures the events used by Bluetooth classic. One of these events is the "meta-event" for low energy. It is therefore necessary for the host to enable this meta-event by using this command. Second, the LE Set Event Mask command is used to enable any low energy events that are required.

8.4.4 Read Buffer Sizes

As explained in the data flow control section, the controller has up to two sets of buffers it uses to buffer data that is transmitted from the host to the controller before it is sent to a peer device.

To read the buffers available in the controller, the host sends both the LE Read Buffer Size command and the Read Buffer Size command, as illustrated in Figure 8–7. Once the Command Complete events have been returned, the host can

8.4 Controller Setup

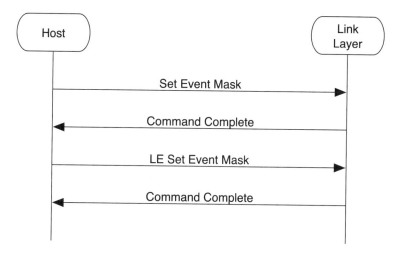

Figure 8–6 HCI set event mask

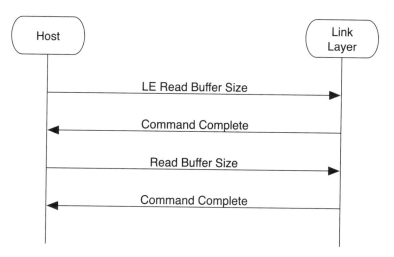

Figure 8–7 Reading the buffer size

determine how many buffers are available to send low energy data packets to connected devices.

8.4.5 Read Supported Features

Another way to ensure that a host can be forward-compatible with a controller is by allowing the host to determine what features a controller supports before sending any feature-specific commands to the controller.

To read the supported features that a controller supports, the host sends the LE Read Supported Features command, as illustrated in Figure 8–8. The controller responds with the Command Complete event, which includes the set of features that this controller supports. This should be sent before any feature-specific commands are sent.

8.4.6 Read Supported States

The controller can be very simple or it can be very complex. So that a host can scale its operations to the capabilities of the controller without trying possibly invalid combinations of states and receiving back failures, a command is provided that can obtain this information.

To obtain the set of supported state combinations, the host sends the LE Read Supported States command, as depicted in Figure 8–9. The controller responds with a Command Complete event that includes the complete list of possible support states.

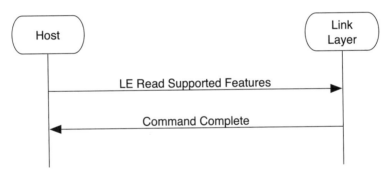

Figure 8–8 Determining what supported features are available on the controller

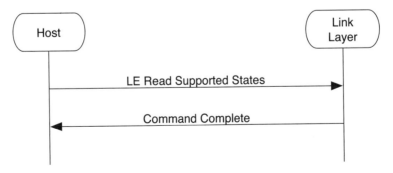

Figure 8–9 Using the LE Read Supported States command

8.4 Controller Setup

The states that can be supported include the following:

- Nonconnectable advertising
- Scannable advertising
- Connectable advertising
- Directed advertising
- Passive scanning
- Active scanning
- Initiating a connection to the master role
- A connection to the slave role

It also has bits that determine if various combinations of states can be supported, such as the following:

- Nonconnectable advertising and passive scanning at the same time
- Nonconnectable advertising and a connection in the slave role at the same time

With this information, the host can determine whether a given command to start advertising, scanning, or initiating will succeed.

8.4.7 Random Numbers

The controller has a very good source of random numbers. Typically, these random numbers come from some physical property of the device; for example, the phase noise between a free running oscillator and an external crystal time source. As with most random number generators that are based off inherently chaotic physical properties, the number of random numbers that can be generated is limited; therefore, the host should ask for as few random numbers as it possibly can. It can then use these random numbers as seeds into a more classic pseudo-random number generator or by infrequently injecting randomness into such a generator.

To ask the controller to generate a random number, the host sends the LE Rand command. The controller responds with the random number in a Command Complete event, as shown in Figure 8–10.

8.4.8 Encrypting Data

It is sometimes useful for the host to encrypt data by using the AES-128 encryption engine that is used by low energy. Given that the controller has to have implemented this encryption engine, it makes sense to provide access to this engine to the host.

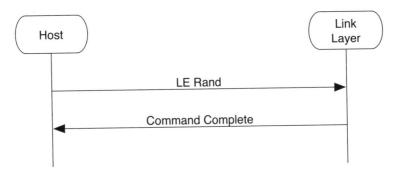

Figure 8-10 Generating random numbers

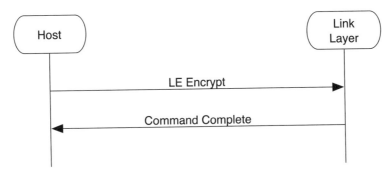

Figure 8-11 The HCI encrypting data

The host can encrypt some plain-text data using an encryption key by sending these pieces of data in an LE Encrypt command to the controller, as illustrated in Figure 8–11. The controller then performs the encryption by using the AES-128 encryption engine and returns the encrypted data in a Command Complete event.

It should be noted that there is no decrypt command. There was a requirement for the host to generate encrypted data for private addresses, but there was no requirement for decrypting data at the host layer. The host can only check that the encrypted data is the same each time something is encrypted and not recover the plain-text data from the encrypted data and an encryption key.

8.4.9 Set Random Address

Some controllers have no fixed device address, or the host wishes to use a private address instead of the fixed address. To accommodate this, the host must program a random address into the controller that can then be used when advertising, active scanning, or initiating connections.

Typically, a random address is generated by using either a random number or a combination of a random number and an identity resolving key. As Figure 8–12 demonstrates, the host requests a good random number from the controller by using

8.4 Controller Setup

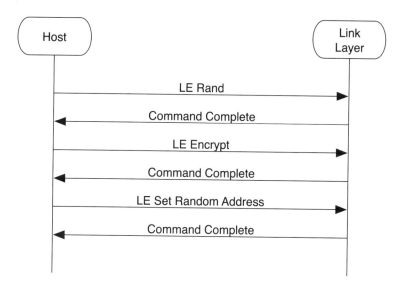

Figure 8-12 Setting a random address

the LE Rand command. Next, it uses that as the plain-text data along with its identity resolving key as inputs into the encryption engine by using the LE Encrypt command. The resultant value can then be used to set the random address via the LE Set Random Address command. The random address is available for use by other commands when the Command Complete event is returned.

8.4.10 White Lists

The controller has a list of device addresses that can be used for white listing devices. This is especially useful when searching for a few known devices in the fog of advertising packets in busy locations. The white list has a fixed size, so the first thing that needs to be determined is how large this list is. The list can then be managed by using commands to clear, add, and remove devices from this list. The controller can use the list, under the control of the host, to filter advertising packets.

To read the white list size, the LE Read White List Size command is sent by the host to the controller (see Figure 8-13). The controller responds with a Command Complete event that contains the maximum number of entries in its white list. The host can clear all the entries in this list by using the LE Clear White List command. It can also add devices to the white list by using the LE Add Device To White List command. It is also possible to remove a single entry in the white list by using the LE Remove Device From White List command.

It should be noted that it is not possible to change the contents of the white list while it is being used. For example, if the device is advertising and using the white list to filter the devices to and from which it will respond to scan requests or

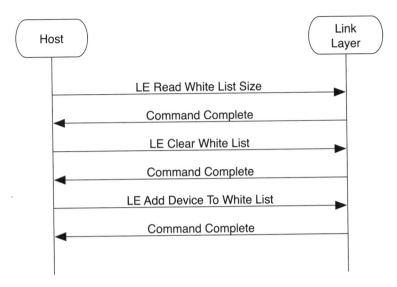

Figure 8–13 Reading a white list

connection requests, the white list cannot be changed until advertising is disabled. The white list can then be changed and advertising re-enabled.

8.5 Broadcasting and Observing

The most primitive form of communication possible between two Bluetooth low energy devices is the broadcasting and observing model. These use advertising and scanning to transmit and receive data.

8.5.1 Advertising

The controller has two sets of data that can be transmitted by using advertising: advertising data and scan response data. It also has a set of parameters that are used for determining how and when it should transmit advertising packets.

The LE Set Advertising Parameters command allows the host to configure the advertising parameters (see Figure 8–14). This includes the minimum and maximum interval that the host requires the controller to advertise, anywhere from 20 milliseconds to 10.24 seconds. The type of advertising is also specified here. Four types of advertising are available:

- Connectable undirected advertising that is used for general advertising of the advertising and scan response data. This allows any other device to connect to this device.

8.5 Broadcasting and Observing

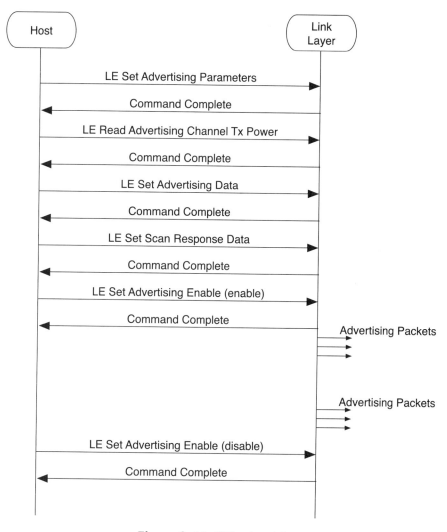

Figure 8–14 HCI advertising

- Connectable directed advertising that is used to request a particular peer device to connect. This does not include any advertising data.
- Scannable undirected advertising that is used to broadcast advertising data and scan response data to active scanners.
- Nonconnectable undirected advertising that is used to just broadcast advertising data.

Another parameter of the LE Set Advertising Parameters command is the address type to be used in the advertising data. This can either be the fixed device address or the random address just described. If the type of advertising used is directed advertising, the peer device address is also included. The final two parameters are the advertising channel map that is used to determine which of the three advertising channels should be used and the advertising filter policy. The filter policy determines if the white list is used to help filter out advertising packets that are received in response to an advertising packet. This can be set to one of the following states:

- Allow a scan request or connect request from any device.
- Allow a scan request only from devices in the white list, but allow connect requests from any device.
- Allow a scan request from any device, but only allow connect requests from devices in the white list.
- Allow only scan requests and connect requests from devices in the white list.

The next command that can be sent is the LE Read Advertising Channel Tx Power command. This returns the transmit power used when advertising. This is useful because it allows this value to be included in the advertising data or the scan response data to enable proximity pairing, or sorting devices by path loss in a user interface.

The host can then configure the advertising data and scan response data by using the LE Set Advertising Data and the LE Set Scan Response Data commands.

Once everything is configured, advertising can be enabled or disabled by using the LE Set Advertising Enable command. While advertising is enabled, the controller advertises by using the configured parameters.

8.5.2 Passive Scanning

To receive advertising data from peer devices, you can use passive scanning.

As illustrated in Figure 8–15, the LE Set Scan Parameters command is used to configure the controller's scanning parameters. These parameters include the scanning filter policy as well as the following:

- **Scan type**—Either passive scanning or active scanning
- **Scan interval**—How often the controller should scan
- **Scan window**—How long the controller should scan for each scan interval
- **Scanning filter policy**—Accept all advertising packets or only those in the white list

The scan interval and window are recommendations from the host to the controller that determine how often and for how long the controller should scan. The interval divided by window determines the duty cycle for which the controller should scan.

8.5 Broadcasting and Observing

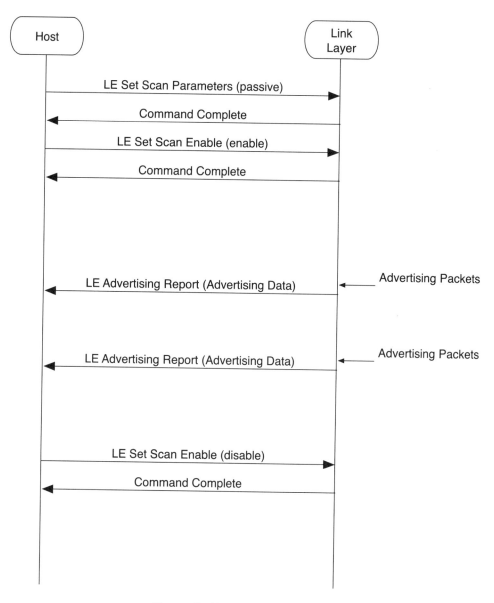

Figure 8–15 HCI passive scanning

For example, if the interval is 100 milliseconds and the window is 10 milliseconds, then the controller should scan for 10 percent of the time. The slowest practical duty cycle that can still pick up a directed advertising packet directed at the controller is a 3.75-millisecond window every 1 second—a 0.4 percent duty cycle.

It is possible to have the interval and window set to the same value. In this case, the scanning is continuous, with a change of scanning frequency once every interval. For example, setting the default parameters at a 10-millisecond interval and a 10-millisecond window results in a 100 percent duty cycle that changes the scanning channel every 10 milliseconds.

The scanning filter policy determines if the white list is used or not. This can allow either all advertising packets to be received, or only advertising packets from devices in the white list to be processed. It should be noted that directed advertising packets that are not addressed to this device would also be discarded, even if sent from a device in the white list.

Once the scanning parameters have been configured, scanning can be enabled by using the LE Set Scan Enable command. Once scanning, the controller sends events up for any advertising packets received that pass the scanning filter policy and other rules. This advertising data is sent to the host in an LE Advertising Report event. In addition to the device address of the advertiser, this event includes any advertising data that was in the packet and the received signal strength indication of this advertising packet. The signal strength can be combined with the transmit power data included in the advertising packet to determine the path-loss of the signal and hence give an approximation of range.

The host can also end scanning by using the same LE Set Scan Enable command but with a parameter set to disable.

8.5.3 Active Scanning

The next complexity up from passive scanning is active scanning. This is used not only to obtain advertising data from peer devices but also their scan response data if possible.

The commands used to configure and enable active scanning are identical to those for passive scanning. However, because the controller will need to send SCAN_REQ packets to the peer device to obtain the scan response data, these packets need to include a device address. Therefore, you use an additional parameter in the LE Set Scan Parameters command to determine whether these Link Layer packets use the fixed device address or the programmed random address, as illustrated in Figure 8-16.

When the controller receives the SCAN_RSP packets, it sends the same LE Advertising Report event to the host. This event also includes the advertising packet type of the Link Layer packet that caused this event. Thus, it's possible for the host to distinguish between a peer device that is advertising as connectable or nonconnectable, or whether the device can be scanned, or if this data is advertising data or scan response data.

8.6 Initiating Connections

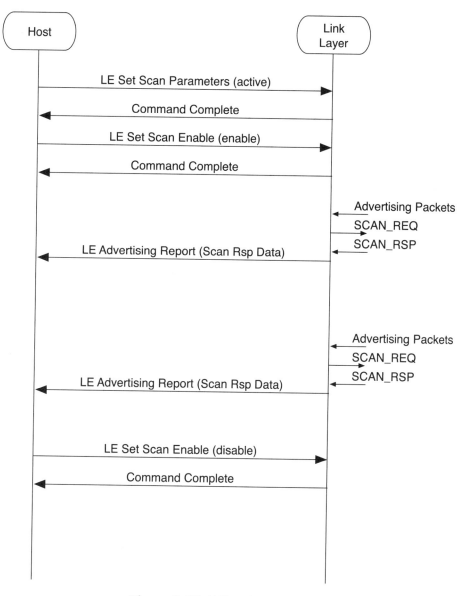

Figure 8–16 HCI active scanning

8.6 Initiating Connections

Advertising and scanning is only useful up to a point. To perform most application work, a connection is required between two devices. To set up a connection, one device must be advertising as connectable, and the other device must be initiating. It is possible to initiate a connection to either a white list or directly to a device. Creating a connection can take some time; therefore, it is useful to be able to cancel the connection creation procedure if the user or application decides that it's no longer necessary.

8.6.1 Initiating Connection to White List

The most common way to connect to a device is for the host to add the device to the white list and then initiate a connection to the white list. By doing this, a controller can initiate a connection to many devices at the same time. Effectively, this allows a host to ask the controller to initiate a connection to device A, B, C, D, E, F, and so on, all at the same time.

To initiate a connection to one or more devices in the white list, the white list must include that device. As depicted in Figure 8–17, the host uses the LE Add Device To White List and other white list management commands to do this. Once the host is happy with the set of device addresses in the white list, it sends the LE Create Connection command to the controller.

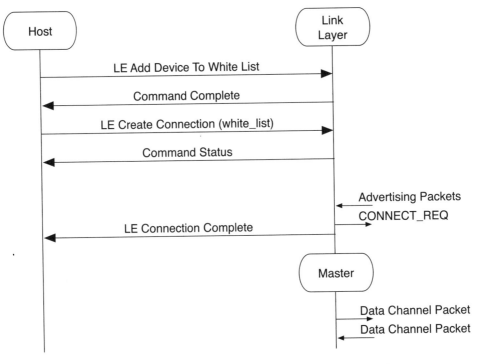

Figure 8–17 HCI initiating a connection to a white list

8.6 Initiating Connections

The LE Create Connection command includes a number of parameters, including the following:

- **Scan interval and scan window**—Just as with the passive scanning parameters, these determine how often the controller should listen for advertising packets from devices.

- **Initiator filter policy**—This should be set to "use white list" to initiate a connection to any device in the white list.

- **Initiating address type**—This determines whether the fixed device address or random address is used in the CONNECT_REQ packet.

- **Initial connection parameters**—This is used to determine how often the master transmits to the slave and the latency that the slave is allowed to use by ignoring the master, as well as the supervision timeout and the expected quantity of data that is to be sent to or from the slave at each connection interval

It should be noted that the initial connection parameters will be identical for each device in the white list. This command is therefore very useful when connecting to lots of similar devices—for example, automation sensors—but not that useful when connecting to many diverse types of devices. For those circumstances, the host should use the worst-case connection parameters in the LE Create Connection command.

If an advertising packet is received from a device that is in the white list, and this advertising packet is a connectable advertising packet type, the controller will send the CONNECT_REQ packet with all the information required and then generate the LE Connection Complete event. The peer device will also send the LE Connection Complete event up to its host once it receives the CONNECT_REQ packet.

The LE Connection Complete event includes the connection handle that is used to label data packets sent from the host to the controller and the controller to the host for this connection. This event also includes the current role of this controller (either master or slave). It is possible for a device to be advertising as connectable and initiating a connection at the same time; this is useful to determine which one of these succeeded. The event also includes the peer device address as well as the interval, latency, and supervision timeout connection parameters. Finally, the event includes the master's clock accuracy that is needed to determine how much window widening is required on a slave device. This parameter is provided to the host for informational purposes only.

The other side effect of sending the LE Connection Complete event from the controller to the host is that any advertising or initiating that was ongoing while the connection was being created automatically stops. Therefore, if the host wants to initiate connections to other devices or continue advertising, it must issue new commands to the controller to do so.

8.6.2 Initiating a Connection to a Device

It is also possible to initiate a connection to a single specific device.

To initiate a connection with a single device, the host uses the same LE Create Connection command, as demonstrated in Figure 8–18. However, the initiator filter policy is set to ignore the white list, and other parameters are used to define the device address of the peer device to which it is connecting. Apart from these minor differences, the connection procedure is the same as when initiating to a white list; an LE Connection Complete event is generated on both devices when the connection has been created.

8.6.3 Canceling Initiating a Connection

Sometimes a connection is initiated to a device that is not responding, probably because it is nowhere near the initiating device, and the host wants to cancel this connection request to do something else.

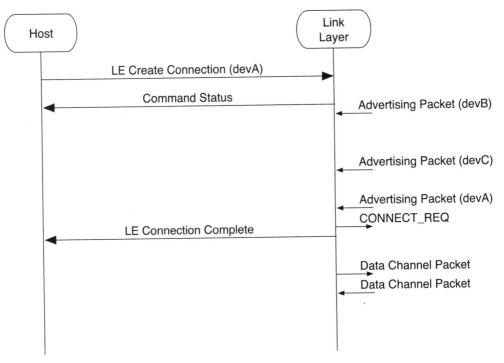

Figure 8–18 HCI initiating a connection to a device

8.6 Initiating Connections

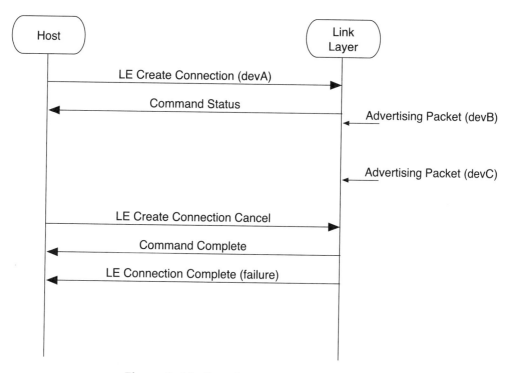

Figure 8–19 Canceling a connection initiation

In this example, which is shown in Figure 8–19, the host uses the LE Create Connection command to attempt to connect to a specific device. This also works for initiating a connection to a white list. When the host wants to cancel the initiating of this connection, it sends an LE Create Connection Cancel command to the controller that responds with a Command Complete event for the cancel command as well as an LE Connection Complete event for the initial LE Create Connection command. It is important to complete all commands, and the LE Create Connection command is completed with the LE Connection Complete event, regardless of whether the connection was created.

It should be noted that there is a race condition here. It is possible for the host to send the LE Create Connection Cancel command to the controller at approximately the same time that the controller sends a CONNECT_REQ packet to the peer device but before the LE Connection Complete event is sent to the host to notify it of the new connection. In this condition, the LE Create Connection Cancel command will be completed by using the Command Complete event, but the LE Connection Complete event will be sent up to the host reporting the new connection that has been created. It is possible, therefore, to have created a valid connection even when trying to cancel this very same connection.

8.7 Connection Management

Once a connection has been created, devices can start to manage that connection to lower power consumption, increase or decrease latency, start encryption, or ultimately terminate the connection.

8.7.1 Connection Update

If the connection parameters that are currently being used on a given connection are no longer useful, the host of the master can change the connection parameters. This could be because the connection was initially created with a very fast connection interval to help the devices initially configure themselves, but once the services are in use, a much longer connection interval is much more beneficial to save power.

The master can change the connection parameters by using the LE Connection Update command, as shown in Figure 8–20. This includes the requested new values for the connection interval and connection latency as well as the supervision timeout and new expected connection event length.

The controller responds with a Command Status event before sending a Link Layer connection update request packet to the peer device (for more information on this, go to Chapter 7, Section 7.10.1). This packet includes the timing of an instant when these new connection parameters will take place. Once this instant passes, the new connection parameters are used, and the LE Connection Update Complete event notifies the host that the new connection parameters have been updated.

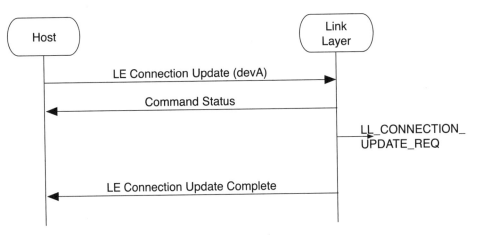

Figure 8–20 Changing the connection parameters

8.7 Connection Management

8.7.2 Channel Map Update

The host might have information about the local channel usage and wish to communicate this information to the controller. For example, it might be co-located with a Wi-Fi radio that is connected to an access point on a given channel and therefore communicating that the low energy channels in the same part of the band would reduce the possibility of the two radios from directly interfering with one another.

There is no way to directly instruct the controller to send a Link Layer channel map request to the peer (for more information about this, go to Chapter 7, Section 7.10.2). However, the host can send the LE Set Host Channel Classification command to the controller, as illustrated in Figure 8-21. This includes a bit field that denotes each Link Layer data channel as either bad or unknown. It is obviously not possible to denote a data channel as good from the point of view of the host because the controller might be measuring the packet error rates by channel already and noticed that some channels denoted as unknown by the host are actually bad.

The LE Set Host Channel Classification command causes a Command Complete event to be immediately sent to the host. The controller can use the Link Layer control procedures to change the channel map at any time it wants. It is possible for the host to monitor the channel map on a given connection by using the LE Read Channel Map command. This command returns whether each Link Layer data channel is used or unused at this time.

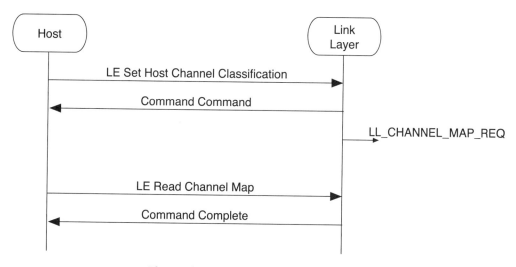

Figure 8-21 Updating the channel map

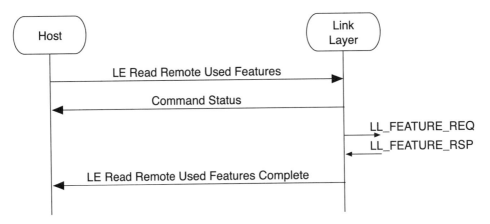

Figure 8–22 Exposing connection features

8.7.3 Feature Exchange

It is possible for the host to discover the features that are available on a connection. For example, even if the local controller supports encryption, encryption can only be used if the peer controller also supports it.

Figure 8–22 shows how the master's host can ask for the remote used features by sending the LE Read Remote Used Features command. This causes a Command Status event to be returned, and the Link Layer feature request (LL_FEATURE_REQ) and response (LL_FEATURE_RSP) to be exchanged (for more information about this, go to Chapter 7, Section 7.10.6). The controller then sends the feature response Link Layer message information to the host in the LE Read Remote Used Features Complete event.

8.7.4 Version Exchange

When debugging devices, especially those that you do not directly control, it is sometimes useful to find the Link layer version information. This information can then be used to help contact the company that manufactured this device so that you can fix the problem. This information can be requested by either the master or slave host, allowing both devices to debug the link if necessary.

To exchange the version information, the host sends the LE Read Remote Version Information command to the controller, as shown in Figure 8–23. The controller responds with a Command Status event and starts the Link Layer version exchange (for more information about this, go to Chapter 7, Section 7.10.5).

Once the version information has been exchanged, the controller sends the LE Read Remote Version Information Complete event to the host with the peer device's version information.

8.7 Connection Management

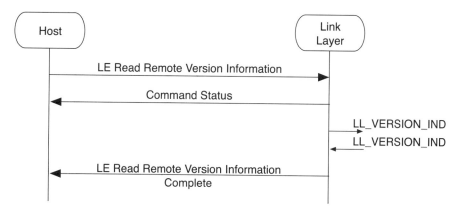

Figure 8–23 Obtaining version information

Note that if the LE Read Remote Version Information command is sent a second time, the same sequence of events is returned, but no Link Layer procedure will be performed because the version information is static information and is cached in the controller. This is also true if the remote device has already exchanged version information before the local host sent the command to the controller. Because the controller already has the version information from the remote device, it immediately sends the completion event after the status event.

8.7.5 Starting Encryption

It is possible for the host to enable the encryption of data packets while in a connection, as long as both sides have a shared secret. This shared secret is set up by the Security Manager, either during the initial paring process or through key distribution during bonding.

Two sequences of commands and events must be considered: one from the master's point of view and one from the slave's point of view.

The master's host can ask for the Link Layer to start encryption by sending the LE Start Encryption command that includes the key used to encrypt the connection (see Figure 8–24). The controller returns a Command Status event while encryption is started. The Link Layer then starts encryption (for more information about this, go to Chapter 7, Section 7.10.3). Once encryption has started, the controller sends the Encryption Change event to the master's host to notify it as to whether encryption is now on or if there was a problem with encryption.

From the point of view of the slave's host, the sequence of events and commands is slightly different. As depicted in Figure 8–25, the first event that warns the slave's host that encryption is being enabled is the LE Long Term Key Request event. This event needs to be responded to, by the host, by sending the LE Long Term

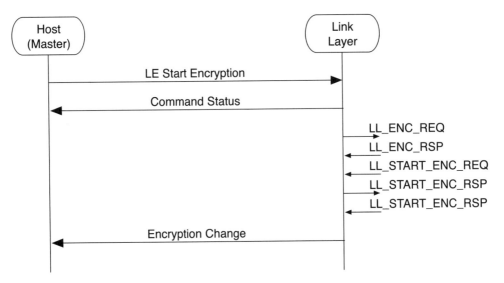

Figure 8–24 HCI master starting encryption

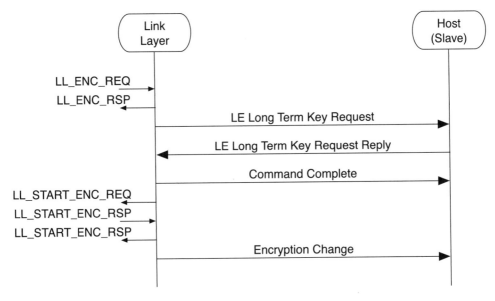

Figure 8–25 HCI slave starting encryption

8.7 Connection Management

Key Request Reply command that includes the key to be used for encrypting the connection. Because this is a command, a Command Complete event is used to complete it. After the connection is encrypted, the Encryption Change event is sent to the slave's host to notify it of the new encryption state.

8.7.6 Restarting Encryption

Sometimes, it is necessary for the host to restart encryption, either by using a new encryption key or just to refresh the initialization vector that is generated as part of the starting of encryption.

As is illustrated in Figure 8–26, from the point of view of the master's host, the sequence of commands and events are identical to the starting of encryption. The Link Layer has to send more packets, first to pause encryption and then to restart it.

From the point of view of the slave's host, the events and commands are also identical, as shown in Figure 8–27.

It should be noted that it is not possible to turn off encryption in low energy and then send application or host data. Therefore, it is not necessary to send an Encryption Key Refresh Complete as soon as encryption is paused.

This is because of a problem discovered in early versions of Bluetooth classic. Hosts that desired maximum security would try to refresh keys regularly to make

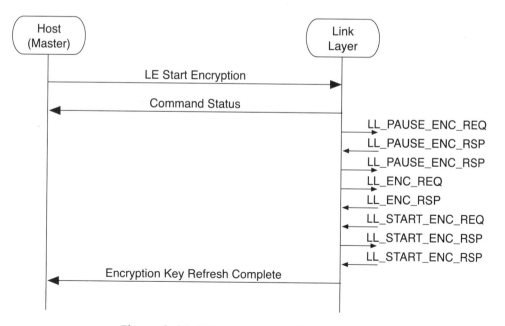

Figure 8–26 HCI master restarting encryption

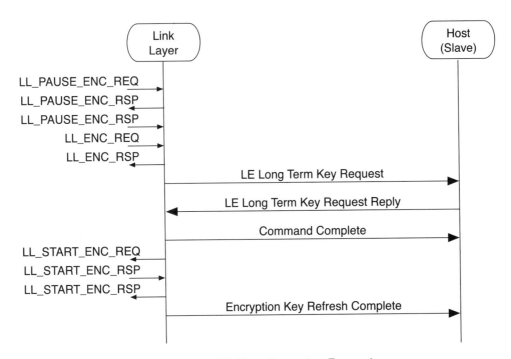

Figure 8-27 HCI Slave Restarting Encryption

them more difficult to crack. However, refreshing keys required turning off encryption, but while encryption was off, data could still be sent. The host couldn't get around this by simply pausing its flow of data to the controller because there might be data in the controller's buffers awaiting transmission. This meant that in early versions of Bluetooth classic, hosts trying to increase security ended up actually decreasing security and causing the controller to send unencrypted data. This was fixed in a later version of Bluetooth classic.

Once a host has requested encryption on a connection, there is usually no reason to disable it. So when encryption is paused, data transmission is also paused, closing the security hole that existed in early Bluetooth specifications.

8.7.7 Terminating a Connection

Once either host has decided that no more data needs to be sent on a connection, or if maintaining the connection would use more power than disconnecting and reconnecting at a later time, then the host can terminate the connection.

To terminate the link, the host sends a Disconnect command to the Link Layer, as shown in Figure 8-28. The Link Layer responds with a Command Status and then attempts to terminate the link by using the Link Layer procedures, as defined

8.7 Connection Management

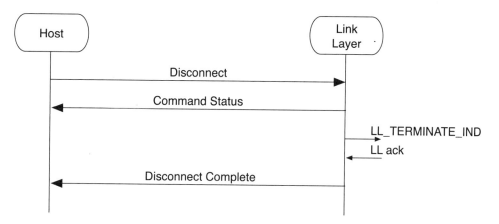

Figure 8-28 Terminating the connection

in Chapter 7, Section 7.10.7. Once the link is terminated, the controller sends the Disconnect Complete event to the host.

Is it also possible for the host to receive the Disconnect Complete event at any time when the link terminates due to a supervision timeout or due to an encrypted message integrity check failure. Supervision timeout typically occurs when two devices have moved so far apart from one another that they can no longer receive Link Layer packets from the peer device. Encrypted message integrity check failures should never occur, except when an attacker is attempting to take over a connection, or in very exceptional cases when a very rare pattern of bit errors passes the cyclic redundancy check but fails the message integrity check.

Part III

Host

Chapter 9, Logical Link Control and Adaptation Protocol, describes the multiplexing layer.

Chapter 10, Attributes, explains how a stateless protocol can be used to obtain the state of a device efficiently.

Chapter 11, Security, is the main body of protocol that is used to provide security services.

Chapter 12, The Generic Access Profile, ties all the pieces of the puzzle together into a high level abstraction that application writers can use.

Chapter 9

Logical Link Control and Adaptation Protocol

> *Anyone who considers protocol unimportant*
> *has never dealt with a cat.*
> —Robert A. Heinlein

The Logical Link Control and Adaptation Protocol (L2CAP) is a protocol multiplexing layer that enables Bluetooth low energy to multiplex three different channels. It also enables segmentation and reassembly of packets that are larger than the underlying radio can deliver. On a Bluetooth classic radio, the L2CAP layer also performs many additional, complicated operations.

9.1 Background

One of the basic concepts for Bluetooth low energy is a radically different connectionless model; this means that you only have to create a connection when you need to send data, and the device can always disconnect at any time. To achieve this, the connectionless model must be extended up to the L2CAP layer; thus, only fixed channels are supported. Fixed channels don't have any configuration parameters to negotiate, and they exist as soon as the lower layers have established a connection; consequently, there is no time wasted waiting for the channel to be created.

When Bluetooth low energy was first designed, it did not use L2CAP. Previously, a Protocol Adaptation Layer (PAL), was designed to be a highly optimized, and severely restrictive multiplexer between two protocols. The PAL looked like the Attribute Protocol and a signaling layer. This was bad for two reasons: flexibility and legacy implementations.

The PAL could only support two types of packet: a single higher-layer protocol or its own signaling layer. There was no segmentation or reassembly, nor was there the ability to separate different protocols. One of the basic design tenets of protocol design is that you layer protocols; each protocol is self-contained. This means that is possible to design, for example, the Security Manager with all the other parts of the system. At the point of implementation, each protocol is a separate layer that can be individually tested. Therefore, the PAL broke this simple rule. The part that killed this approach, however, was not the design, but the lack of flexibility.

Most multiplexing layers perform segmentation and reassembly. This means that a large protocol packet from a higher layer can be segmented into multiple smaller packets appropriately labeled so that they can be transmitted through a system that has packet length restrictions. A good example of this is an ATM network for which each packet is restricted to just a few bytes of data, allowing the rapid switching between different streams. This facilitates the delivery of low-latency audio traffic and bulk data at the same time.

The Host Controller Interface (HCI) supports segmentation and reassembly by using the "start" and "continuation" bits on each data packet. However, the PAL didn't support such a basic feature. This meant that the maximum size of any application data in this layer would be limited to just 24 bytes of data. This severe restriction was the eventual downfall of the PAL.

When L2CAP was proposed as an alternative, the group designing Bluetooth low energy split down the middle: the companies that already had existing Bluetooth implementations and the companies that didn't. In some standards bodies, this would have meant many months of acrimonious voting to attempt to force division; this is also typically associated with disruptive political actions like trying to stuff the room with voting members to try to sway the vote one way or the other. In Bluetooth, this is not the standard approach. Instead, a paper on the various costs of each approach was written showing the cost of adding L2CAP. The deciding argument was that the battery life of a device that reported something once a second was reduced from 3.3 years to 3.2 years. So L2CAP did reduce the battery life of the device, but compared with the 7 bytes before the payload of the packet, and the 3 bytes of cyclic redundancy check (CRC) on every single packet whether it was carrying data or not, it was not a significant reduction. This is another example of the attention to detail that the designers of Bluetooth low energy took to consider the system design issues of all the decisions.

L2CAP gives you the ability to plug Bluetooth low energy into an existing L2CAP implementation. It also supports the full segmentation and reassembly from Bluetooth classic, effectively allowing packet sizes of up to 65,535 bytes in length; even though there are no protocols that support this packet size that can be run on Bluetooth low energy. L2CAP also retains the channel model that Bluetooth classic uses.

In Bluetooth classic, the channels come in two different flavors: fixed and connection-oriented. A fixed channel exists for as long as the two devices are connected. These are used primarily for signaling channels, either for basic L2CAP signaling commands or, in v2.0 and later, an Alternate MAC/PHY signaling channel.

Connection-oriented channels can be created at any time by sending a few L2CAP signaling commands to a peer device.

In Bluetooth classic, connection-oriented channels allow data from an individual pair of applications to be considered as separate from the data of other channels. For example, even though connection-oriented channels can add additional data integrity checking, they might have a different flow specification, or they might be a streaming channel rather than a best-effort channel. Connection-oriented channels are great when you have a complex system that has multiple, varied types of data being transmitted at the same time. For example, a phone and a car can have multiple different protocols running at the same time: one stream for the high-quality audio from the phone to the car stereo; one stream for the hands-free operation; another stream for the phone book; and perhaps another stream for an Internet connection.

Opening connection-oriented channels can be a complex operation. Each L2CAP channel has a large number of configuration parameters; seven in the latest specification. This means that in addition to the two messages that have to be exchanged to request a connection to be established, each of the configuration parameters has to be agreed upon before any data is allowed to be sent. This could be fairly quick—just another four messages—or it could be a fairly lengthy operation of proposed values and counter proposals. The other complexity that connection-oriented channels bring is that once they are all configured and data is flowing, a device can renegotiate different parameters. All this increases the latency of the data connection at the expense of more flexibility. For most Bluetooth classic protocols and profiles, this is an acceptable cost because these connections are kept alive for long periods of time.

9.2 L2CAP Channels

In L2CAP, there is a simple concept of a channel. L2CAP, after all, is a multiplexing layer, and to do this, it has multiple channels. A channel is a single sequence of packets, from and to a single pair of services on a single device. Between two devices, there can be multiple channels active at the same time.

In Bluetooth low energy, only fixed channels are supported. A fixed channel is a channel that exists as soon as the two devices are connected; there is no configuration requirement for fixed channels. The future-proofed flexibility still exists to add connection-oriented channels if they are considered necessary.

Table 9–1 presents the L2CAP channel identifiers. Each channel identifier in Bluetooth is a 16-bit number. The channel identifier 0x0000 is reserved and should never be used. Channel identifier 0x0001 is a fixed channel for Bluetooth classic signaling.

Table 9–1 L2CAP Channel Identifiers

Channel Identifier	Use
0x0000	Reserved: cannot be used
0x0001	Bluetooth Classic signaling channel
0x0002	"Connectionless" channel
0x0003	AMP manager protocol
0x0004	Attribute Protocol
0x0005	LE signaling channel
0x0006	Security Manager Protocol
0x0007 to 0x003E	Reserved: may be used in the future
0x003F	AMP test protocol
0x0040 to 0xFFFF	Connection-oriented channels

Channel identifier 0x0002 is a fixed channel used for "connectionless data," although there is no profile that currently uses this. Channel identifier 0x0003 is used for the Alternate MAC/PHY protocol when sending data at high speed is required. Channel identifier 0x003F is used for a test channel for the Alternate MAC/PHY controllers.

There are three Bluetooth low energy channels: Channel identifier 0x0004 is used for the Attribute Protocol (for more information on this, go to Chapter 10, Attributes); Channel identifier 0x0005 is used for the Bluetooth low energy signaling channel; Channel identifier 0x0006 is used for the Security Manager (for more information on this, go to Chapter 11, Security). All the other channel identifiers from 0x0007 to 0x003E are reserved, and channel identifiers from 0x0040 to 0xFFFF can be used for connection-oriented channels.

9.3 The L2CAP Packet Structure

Each L2CAP packet contains a 32-bit header followed by its payload. It is assumed that segmentation and reassembly is used; thus, the length of the packet must be included in the packet header so that the end of the packet can be determined. The segmentation and reassembly scheme used requires the marking of packets over the HCI interface (for more information on this, go to Chapter 8, The Host Controller Interface) as well as within each transmitted packet as either a start or continuation packet. There is no way to denote that a given L2CAP packet segment is the end of the current packet. This means that the only way to determine if the current packet is complete is to either send a new packet, assuming that one is ready to be sent, or to include the packet length in the very first packet sent.

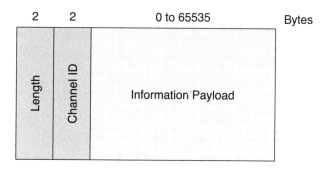

Figure 9-1 L2CAP Packet Structure

As shown in Figure 9-1, the header contains a 2-byte length field followed by the 2-byte channel identifier. This is followed by length bytes of information payload. In Bluetooth classic, the information payload can also include additional headers and information, but in Bluetooth low energy, there are no other structures of significance at the L2CAP layer.

For all Bluetooth low energy channels, the information payload starts with a Maximum Transmission Unit (MTU) size of 23 bytes. MTU is the largest possible size for the information payload in a given L2CAP channel. This means that all Bluetooth low energy devices must support 27-byte packets over the air—4 bytes of the L2CAP header and 23 bytes for the information payload.

9.4 The LE Signaling Channel

The LE signaling channel is used for signaling at the host level. As illustrated in Figure 9-2, each LE signaling channel packet contains a single opcode, followed by any parameters. The following command opcodes are supported on the LE signaling channel:

- Command Reject
- Connection Parameter Update Request
- Connection Parameter Update Response

Whenever a signaling command is sent, an identifier is included in the information payload. This identifier is just 1 byte in length and is used to match responses with requests. For example, if a request was sent with the identifier 0x35, any response that also had the same identifier 0x35 would be the response for that request. This allows multiple requests to be outstanding at the same time, with each request having

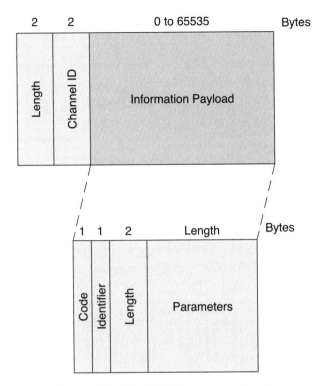

Figure 9-2 The L2CAP command packet

a different identifier. Identifiers can't be reused unless all other identifiers have been used. This leads implementations to use an increment operation to ensure this rule is met. There is just one exception to this: An identifier with the value 0x00 is never used. A side effect of the use of identifiers is that duplicate commands can be silently dropped. This would be useful if the command channels were unreliable, but they are always sent on a reliable bearer, so this rule is rarely invoked.

In Bluetooth low energy, because only one request has been defined, and because this request can only be sent when no other request is outstanding, the logic for identifiers is very simple.

9.4.1 Command Reject

The command reject command is used to reject any nonsupported message that was received by the device. This command is identical to the Bluetooth classic command reject command. It contains a reason code and can contain some data. The reason code can be either *Command not understood* or *Signaling MTU exceeded*.

The *Command not understood* reason code is used when a command was sent to the device that it does not support. This should be sent even for command codes that are not defined at the moment; this allows a device to be forward-compatible with future versions of the specifications.

The *Signaling MTU exceeded* reason code is used when a command is received that is longer than 23 bytes. The default MTU for the signaling channel is just 23 bytes, so if a command were received that was 24 bytes or more, the command reject would be sent in reply.

In Bluetooth classic, another reason code is defined, *Invalid CID in request*, but because no commands are defined that use a channel identifier in Bluetooth low energy, this reason code has never been used.

9.4.2 Connection Parameter Update Request and Response

The connection parameter update request command provides the slave device with the ability to ask for the Link Layer connection parameters to be updated, as demonstrated in Figure 9–3. These parameters include how often the slave wants the master to allow the slave to transmit, the connection event interval, and often the slave wants to be able to ignore the master, the slave latency, and the supervision timeout.

This command would be used when the slave is in a connection for which it wants to modify current connection parameters. For example, the connection event interval might be too fast and therefore wasting too much power. This would not be a problem if the slave latency were reasonably high, but if this is not true, then the slave would have to listen very frequently. Sometimes this is useful, for example, when the devices are first bonding and sending many messages between one another, discovering the services and characteristics of the device. But many other times, having the ability to minimize the number of connection events when the slave has to listen is vitally important for efficient battery life.

This command is only usefully sent from the slave to the master; the master can always initiate a Link Layer connection parameter update control procedure at any time (see Section 7.10.1 in Chapter 7). If the command is sent by the master, the slave would consider it an error and would respond with a Command Reject command with the reason code *Command not understood*.

The slave can send this command at any time. If the master receives the message and can change the connection parameters, it will respond with a Connection Parameter Update Response with a result code set to *accepted*. The master will also initiate the Link Layer connection parameter update control procedure.

Of course, this is just a request, and if the master doesn't like the parameters that the slave wanted, it can reject the request by sending a Connection Parameter Update Response with the result code set to *rejected*. The slave then has two options: accept that the master wants or needs the connection parameters that it is currently using, or terminate the connection. Terminating the connection might appear at first

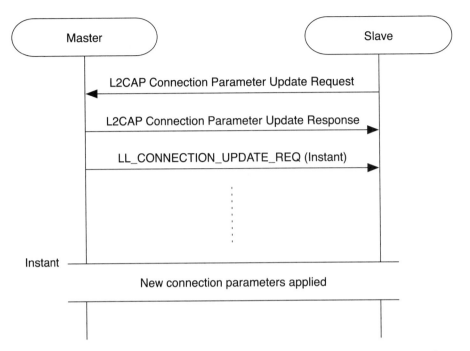

Figure 9-3 The L2CAP Connection Parameter Update Request command

glance to be a fairly drastic approach, but if the slave would burn through its battery in a week with the current connection parameters but would last for years with its requested connection parameters, it might have only one logical choice available.

To reduce the probability of having the master reject the connection parameters from the slave, the slave can request a range of connection event intervals that would be acceptable. A well-designed slave would willingly accept a wide range of intervals. A master device might also be doing some other activities such as a low latency conversational audio connection or a high-quality audio connection and is therefore severely restricted in the range of connection intervals that it can accept. The set of intervals it can accept might be different depending on what it is currently doing, so it might not be the same as the last time the two devices connected.

Another way to increase the chance that the master will accept the connection parameters is to have a reasonably sized slave latency. The master can then choose the most suitable connection event interval, and the slave can then use a slave latency that gives it the best power consumption. For example, if the slave wants to synchronize every 600 milliseconds, it could request a connection interval range of between 100 milliseconds and 750 milliseconds, with a slave latency of 5. If the master chooses 100 milliseconds, the slave could synchronize every 6 connection events. If the master chooses 200 milliseconds, then the slave could ignore 2 out of every 3 connection events, achieving its desired synchronization interval of 600 milliseconds. If the master chooses 300 milliseconds, the slave could ignore every other connection event. If the master chooses 400 milliseconds, the slave could synchronize every 400 milliseconds.

Chapter 10
Attributes

*Data is a precious thing and will last
longer than the systems themselves.*
—*Tim Berners-Lee*

*Civilization advances by extending the
number of important operations which we
can perform without thinking of them.*
—*Alfred North Whitehead*

There are two layers that will be considered in this chapter: the Attribute Protocol Layer and the Generic Attribute Profile Layer. Both are so closely related that it is useful to discuss them at the same time. When Bluetooth low energy was created in the Bluetooth Special Interest Group (SIG), the concepts behind Attribute Protocol were originally created within a non-core working group before being integrated into the core specification. However, at the time of integration an architectural decision was made to split the document into an abstract protocol and a generic profile. Although this is a useful abstraction to make from the specification point of view, it is not useful when attempting to understand how attributes work. The abstraction of generic attribute profile away from the attribute protocol can theoretically allow other generic profiles to be placed above the attribute protocol. And although this is possible, it's not something that is being considered currently.

10.1 Background

When Bluetooth low energy was first designed, there was a big question about what protocol to use. The protocol had to be very simple because any complexity would increase the cost and memory requirements for that protocol. It was also desirable to use the minimum number of protocols as possible. As a result, it was considered that using a single protocol for everything would be the best initial approach. This goal was not entirely met; Bluetooth low energy uses three protocols: Logical Link Control and Adaptation Protocol (L2CAP), Security Manager Protocol (SM), and Attribute Protocol (AP).

The goal was to reduce the number of protocols to a minimum, and each and every service above the Generic Attribute Profile (GATT), including the Generic Access Profile for name and appearance discovery, uses the AP. This allows additional services to be created, built on top of the GATT, for minimal additional cost.

10.1.1 Protocol Proliferation Is Wrong

You might be questioning why protocols are such a bad thing. The whole of computing, and in some senses the rest of the world, revolves around protocols. Most activities have their own protocol: to download a Web page, the Hypertext Transfer Protocol (HTTP) is used; to transfer a file, the File Transfer Protocol (FTP) is used; to log in to another computer securely, the Secure-Shell protocol (SSH) is used. Each protocol is optimized for its own application area. It is not efficient to transmit a large group of files by using HTTP, and it is not efficient to log in to computers by using FTP.

The big difference between Bluetooth low energy and the plethora of Internet protocols is that Bluetooth low energy is not trying to transfer such a wide range of data types. Given that it is not about transferring large quantities of data or streaming music, a single protocol can be designed that only has to deal with the limited set of data types that Bluetooth low energy targets. This protocol is called the Attribute Protocol; it is the foundation and building block for the whole of Bluetooth. To understand Attribute Protocol is to understand Bluetooth low energy.

10.1.2 Data, Data, Everywhere...

When Bluetooth low energy was first discussed, it was clear that as any communications system, it is all about data. Lots of things have data, and Bluetooth low energy is a means by which lots of other devices can access and use this data. This data could be anything: the signal strength of your mobile phone; the state of the battery in your toys; your weight; how many times you've opened the fridge today; how far you've bicycled this morning; what time it is; how much talk time you have on your headset; the latest news headline; who has just sent you a text message; if the chair is being sat upon at the moment; who is in the meeting room; how long you've spoken on your phone this month—anything!

As Figure 10–1 demonstrates, a Bluetooth proximity device might expose its transmit power level, which is an alert level used to notify the user when the connection is lost. It might also have a device name so that the user can more easily identify

10.1 Background

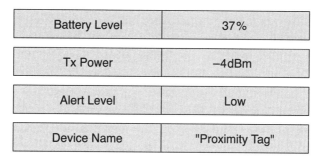

Figure 10–1 Some examples of the types of data that Bluetooth low energy devices might have

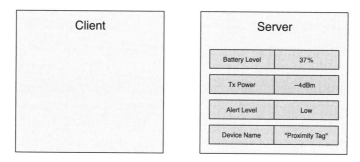

Figure 10–2 Servers have data; clients use this data

the device. Finally, because the device is battery powered, it might also expose its current battery level.

The important concept about data that you need to understand is that some devices have it, and other devices want to use it. In Bluetooth low energy, this distinction is very important because it determines which devices are considered to be servers and which are clients. A device that has data is a server; a device that is using the data from another device is a client. This relationship is illustrated in Figure 10–2.

10.1.3 Data and State

There is another important concept that you need to understand. There is a significant difference between *data* and *state*. Data is a value that represents something, such as a fact or a measurement. Data could be the temperature of the room as measured by the thermometer, or it could be temperature of the room as read by the heating system; thus multiple devices can "know" data. State is a value that represents the status or condition of a device: what it is doing, how it is operating. This state is only known on one device; one device is said to hold this state information. The thermostat measures the room temperature and is therefore said to reflect the state of the temperature for that room.

In this book, "state" refers to the information (data) that resides on the server; "data" refers to that information (again, the data) as it is in transit from the server to the client or held on the client. So, a server is a device that holds a collection of state information. A client is a device that reads or writes this state information, perhaps caching it locally as data. The data on the client is not authoritative because the server's state could have changed since the client last received data from it. When reading the following sections, remember that devices have state, and that the state will be on the server.

10.1.4 Kinds of State

Bluetooth low energy uses three different kinds of state: external, internal, and abstract.

Current physical measurements represent the state of a physical sensor or similar interface. For example, let's consider a bathroom scale. As shown in Figure 10–3, measurements for this device might include the current temperature of the room, the current battery state of the weighing scales, or the weight of the person who last used the scale. These are all known as *external state*; state that every time you read it might result in a different value because it is being measured by using an external sensor.

The next type of state is *internal state* (see Figure 10–4). Some devices have *state machines* that represent their current internal state. They don't represent the external state of a sensor, but how the device is currently functioning. This could include things such as the state of the call on the phone, whether time is currently being synchronized by using a GPS receiver, or if the light is still changing brightness due to an earlier dimming command.

The last type of state is an *abstract state* (see Figure 10–5). This is state information that is only relevant at a momentary point in time; it does not represent the

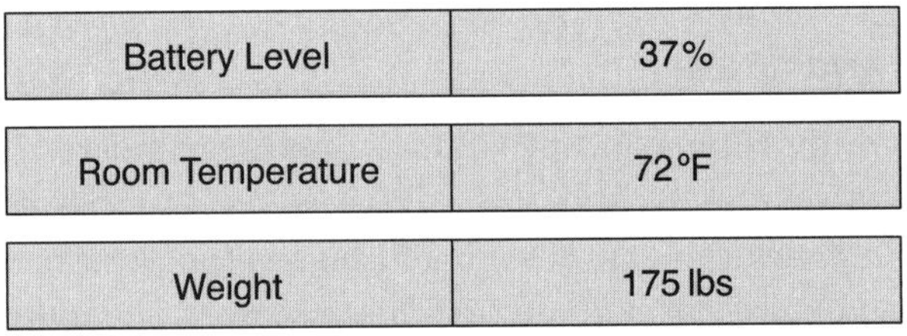

Battery Level	37%
Room Temperature	72°F
Weight	175 lbs

Figure 10–3 Physical measurements

10.1 Background

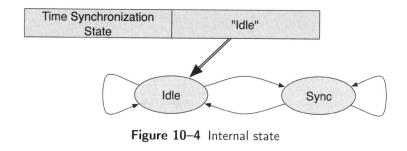

Figure 10–4 Internal state

Figure 10–5 Abstract state

current external or internal state of the device. Examples of this type of state include a way to command a light to toggle its on/off state, a way to request a device to immediately alert, or a way for a device to control when time is synchronized and how to cancel an in-progress synchronization. In the Attribute Protocol, these are known as *control points*. Typically, these are attributes that cannot be read; they can be written or notified.

10.1.5 State Machines

The most interesting aspect of the Attribute Protocol and the types of state that can be exposed is that it explicitly supports exposing finite state machines. A state machine represents the internal state for the device. A state machine also has one or more external inputs into the machine. These external inputs are momentary commands that move the state machine from one state to another, as determined by other state information or behavior; this is an abstract state, or control point.

By using the combination of internal state and control points, it is possible to fully expose the workings and behavior of a finite state machine on a device. This is interesting from two points of view. First, by exposing finite state machines, their inputs, and their current state, the behavior of a device can be exposed explicitly. By exposing inputs, other devices can interact with this device. Second, it is possible to define the full behavior of a finite state machine, including invalid behavior. By doing this, any device can send an input on any control point into a state machine, and the behavior defined for that state machine will still define what will happen.

Light State	Input Received	Result
on	turn on	light = on
on	turn off	light = off
on	toggle	light = off
off	turn on	light = on
off	turn off	light = off
off	toggle	light = on

Figure 10–6 An example of state transitions for a light

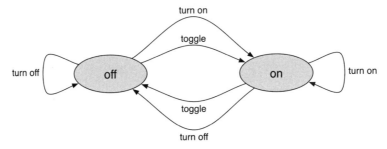

Figure 10–7 An example of a light state machine

Consider for a moment a very simple state machine for a light. The light can be considered to have a finite state machine with two states: on and off, as illustrated in Figures 10–6 and 10–7. It could be possible to read its current state and also write this state to change the light's state. However, it is also useful to consider that this state machine has three possible inputs: *turn on*, *turn off*, and *toggle*. Most of the state machine inputs can map to a valid and logical next state. For example, sending *turn on* to a light that is off will turn it on, sending *turn off* to a light that is on will turn it off. However, sending *turn on* to a light that is on will keep it on even though this might be considered an invalid behavior. Similarly, sending *turn off* to a light that is off will keep it off. Also, sending *toggle* to a light that was on will turn it off. Sending *toggle* to a light that was off will turn it on.

The interesting thing about the *toggle* input is that it significantly reduces the volume of traffic that needs to be sent over the radio to change the light's state. Without exposing this abstract control point, a light switch would need to first read the current light state, toggle this data internally, and then write the new value to the light state. This requires a minimum of three different messages to be sent: a request for the current light state, a response including that light state, and a command to set the light state to a new value. By adding the toggle command to the light's finite state machine, it is possible to remove most of these messages by just commanding the light state machine's abstract control point to accept the *toggle* command. The light switch can then send a single toggle command to the light. The light does the toggling of the light state on the server; the light switch doesn't need to know the old or new state.

Exposing a state machine is therefore more efficient in terms of messages that need to be sent over the radio, but is also more interoperable because it is impossible to command a state machine into a state that has not been defined by the behavior of this state machine. Therefore, by defining all possible states, and the behavior of all inputs in all possible states, an interoperable and optimal protocol can be used.

10.1.6 Services and Profiles

The most interesting architectural change between Bluetooth classic and Bluetooth low energy is the service and profile architecture. In Bluetooth classic, most profiles also include protocols, defined behavior, and interoperability guidelines. These classic profiles are therefore highly complex and encompass many different concepts. The biggest problem is that the profiles define just two types of device, one at either end of the link. The behavior of each device is then explicitly defined. At first glance, this might appear to be a very useful thing to do. If you have a phone and a Bluetooth car kit, it would be very useful to define what each device must do and how it interacts with the other device to enable a given use case. Unfortunately, this has a few problems.

The first problem with existing profiles is that the behavior of a given device in the network is not explicitly defined on its own. This means that even though the behavior of the two devices is defined, it is sometimes not explicitly clear what the behavior of each individual device should be.

This leads to ambiguities wherein each device believes it is the other device's job to carry out an action, and thus the action never gets done. For example, the Hands-Free Profile (HFP) says, "either the HF or the AG shall initiate the establishment of an Audio Connection whenever necessary." So, which device initiates the audio connection, the HF or the AG? What happens if they both attempt to do this at the same time? This is an interoperability nightmare.

The obvious solution to this is to define the behavior of each device separately, to make it explicit what each device should do.

The second problem with existing profiles is that it is almost impossible to use the profile in a way that was not initially envisioned. Because profiles define how the two devices interoperate with one another, it is very difficult to then make it work with a slightly different device. Even within profiles, this becomes difficult. For example, the hands-free profile defines a phone and a car kit, yet the most-implemented use case is a phone with a headset; the phone doesn't know it is talking with a car kit or a headset and therefore continuously sends user interface status updates to the other device because it might be a car kit. This wastes power because a headset really doesn't care about the signal strength going from four bars to three bars. The obvious solution to this is to define the behavior of each device without the need to know the device's functionality.

In Bluetooth low energy, these problems have been tackled by taking a radically different approach. First, because we have a pure client-server architecture, we have separate documents that describe the behavior for a given use case on the server and on a client. The server's behavior is defined in a service specification, whereas the client's behavior is defined in a profile specification. As illustrated in Figure 10-8, this means that the service specification defines the state that is exposed in the server by using an attribute database as well as the behavior that is available through these attributes.

Some attributes on a service might be readable, returning either historical or current data. Some attributes might be writable and make it possible for commands to be sent to the service. Some attributes expose the state of a finite state machine that when combined with control points provide fully exposable behavior. The profile specifications define how to use one or more services to enable a given use case (for example, how to configure the attributes exposed for a service in an attribute database on the server to ask the service do something that client needs it to do).

The main advantage of this split is that the server has a known and defined behavior. It does what it does, as defined by the service specification, without any interest in how the client is using it. This means that the service can be individually unit tested and that it is independent of the client. Any client can use that service if it needs to do so. For example, if there is a time service, this service could be used by one client to obtain the current time; it could be used by another client to read the current time periodically to determine it's own clock drift; it could be used by

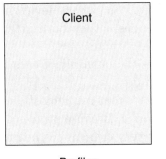
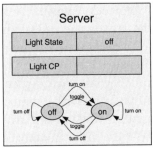

Figure 10-8 The profile/service architecture

10.1 Background

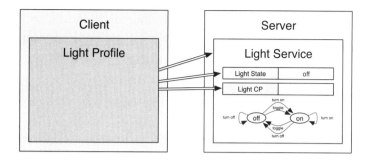

Figure 10-9 An example of a light profile and service

another client to request that a GPS receiver is used to obtain the most accurate time possible. The time service doesn't care what the client is doing; it just does it.

For example, as demonstrated in Figure 10-9, a light can expose a Light Service with two pieces of data: the current physical light state, and the abstract light control point to allow a client to control the state of the light. A light switch would implement the light profile that knows how to find the light service, how to read the current light state, and how to control the light state. The light switch knows the behavior that is exposed because this must be the same for every instance of the light service.

The client is also at an advantage using this system. The profiles are written without defining the behavior of the server—these are defined by the services—and therefore very much simpler. The client profiles are in essence a set of rules for discovering, connecting, configuring, and using services. They also include standard procedures for performing various actions that are required by the client. The clients, and their profiles, can use any combination of services to achieve their goals. For example, a client could combine the use of the time service with a temperature service to allow temperature to be monitored over time, without the need to have a real-time clock in the client and without the need to collect this data in real time.

Take, for example, a home security system that knows that the house is unoccupied, but the homeowners would like the house to appear to be occupied by turning on and off lights randomly. This could be accomplished by defining a new profile that implemented some very simple sets of instructions, as illustrated in the following script:

```
loop forever:
    wait <random period from 10 seconds to 3 hours>
    connect to a light:
        send "toggle" to control point
        disconnect
```

Alternatively, if you were managing an office and wanted to make sure all the lights were off when people were not at work, this could be implemented in yet another profile that is represented by this set of instructions:

```
loop forever:
    wait <until start of next working day>
    connect to lights:
        send "on" to control point
        disconnect

    wait <until end of working day>
    connect to lights:
        send "off" to control point
        disconnect
```

It is this combination of multiple services in the client that is one of the most powerful concepts in Bluetooth low energy. Each individual service can be kept very simple; it is the combination of services that provides the complexity and richness of the system. For this to be true, services must be atomic. Atomic in this context means that the services perform only one set of actions. By making services atomic, they can be reused by multiple clients, all doing different things.

This allows the first problem identified with classic profiles to be solved—it is almost impossible to define a classic profile that has explicit state. By defining the services as separate specifications, it is possible to keep them atomic so that the quantity of behavior in each service is very small. The small quantity of behavior means it is very much easier to define explicitly what a service does, and, therefore, it is also easier to test this by using standard unit testing methodologies. This means that atomic services have explicit state, and that this state can be relied upon by clients.

This approach also solves the second problem identified with classic profiles—it is almost impossible to use a classic profile in a way that was not envisioned. By splitting the use cases into services that are atomic with known behavior, it is possible to define clients that can use these services. Clients can use these either in isolation or in combination with other services. Because the services have explicit behavior, the complexity of combining services is minimized. Also, because each service is atomic, there is no "bleeding" of behavior from one service to another. Each service is separate, and its behavior is not dependent upon the state of another service. This means that services can be combined in any order that is conceivable.

This also means that clients can use services on a device in novel ways. There is no actual need for profiles; it is the services that define how devices interoperate, profiles just define the standard ways that this can be done for a given use case. Therefore,

10.2 Attributes

a device can determine that it can combine not only the temperature service and time service but also the solar panel's power generation service to determine the current weather. This is probably not a profile or use case that would be defined by the Bluetooth SIG, but it could be something that a manufacturer might wish to do. By splitting the behavior of the server into services, it is possible to combine this behavior in interesting, novel, and useful ways in any client.

10.2 Attributes

To understand the Attribute Protocol, you first must understand an attribute. Defined broadly, an attribute is a piece of labeled, addressable data. In the following subsections, we'll look more closely at what this means and how you can use attributes in a practical sense.

10.2.1 Attribute

Figure 10–10 shows that an attribute is composed of three values: the attribute handle, the attribute type, and the attribute value.

10.2.2 The Attribute Handle

A device can contain many attributes. For example, a temperature sensor might contain an attribute for the temperature, one for the device name, and one for the battery state. It could be considered that the attribute type would be sufficient to identify a given attribute; for example, just asking for the temperature should return the temperature, asking for the device name should return the device name, and so on. However, what if the device contains two temperature sensors: an indoor temperature sensor and an outdoor temperature sensor? In this case, you cannot just read the temperature sensor; you need to read the first or second temperature attribute. This problem becomes much more complex when you consider that you could have an arbitrary number of temperature sensors.[1]

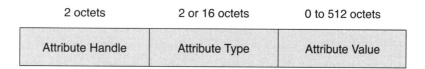

Figure 10–10 The structure of an attribute

---

1. The computer used to write this book has 23 different temperature sensors.

To solve the this problem, instead of addressing attributes by their type, you use a 16-bit address called the *attribute handle*. Valid handles are 0x0001 to 0xFFFF. Handle 0x0000 is an invalid handle and cannot be used to address an attribute. You can consider these handles to be the memory address, port number, or hardware register address for the attribute value, depending on your particular background in software, hardware, or embedded engineering.

10.2.3 Attribute Type

There are many different types of data that can be exposed: temperature, pressure, volume, distance, power, time, charge, Boolean on/off state, state machine states, and so on. The type of the data that is exposed is called the *attribute type*. Given the different possible types of data that can be exposed, a 128-bit number is used to identify the type of the attribute. This unique identifier is known as a *Universally Unique Identifier* (UUID).[2]

UUIDs are huge. A 128-bit UUID requires 16 bytes of data to be sent between devices so that each device can identify the type of the data. To enable the efficient transfer of data types between devices, the Bluetooth SIG has defined a single 128-bit UUID, called the *Bluetooth Base UUID* that can be combined with a small 16-bit number. The use of a defined Bluetooth Base UUID means that this UUID and any derived UUID still follow the rules for allocating UUIDs. It also means that when sending UUIDs between devices for well-known values, only the short version of the UUID can be sent and then recombined with the Bluetooth Base UUID when it is received.

The Bluetooth Base UUID is defined as the following:

$$00000000 - 0000 - 1000 - 8000 - 00805F9B34FB$$

When a short 16-bit Bluetooth UUID is sent, say the value 0x2A01, the full 128-bit UUID would be the following:

$$00002A01 - 0000 - 1000 - 8000 - 00805F9B34FB$$

When referring to these 16-bit Bluetooth UUIDs, the values of the short UUIDs are very rarely used. Instead, a name of these values is used, surrounded by guillemets ("≪" and "≫"). So, for example, the name ≪Includes≫ is a 16-bit Bluetooth UUID that has the value 0x2802. There are many 16-bit Bluetooth UUIDs that are defined. The UUID itself does not define the usage of the UUIDs, but the UUIDs that are

---

[2]. Universally Unique Identifiers are defined in RFC 4122, which is functionally equivalent to ISO/IEC 9834-8.

10.2 Attributes

used by Bluetooth low energy are arranged into the following groups for human readability when debugging:

- 0x1800 through 0x26FF are for Service UUIDs
- 0x2700 through 0x27FF are for Units
- 0x2800 through 0x28FF are for Attribute Types
- 0x2900 through 0x29FF are for Characteristic Descriptors
- 0x2A00 through 0x7FFF are for Characteristic Types

10.2.4 Attribute Value

The state data that a device exposes is available in an *attribute value*. Each attribute has a value that can be any size from 0 bytes to a maximum of 512 bytes in length, although the size is fixed for some attribute types. The value of the attribute is not significant to the Attribute Protocol, but it is significant to the layers above that include the Generic Attribute Profile and GATT-based services and profiles.

10.2.4.1 Service UUIDs

Each service can be identified by using a UUID. This can be either a 16-bit UUID or a full 128-bit UUID. There are 3,840 services that can be allocated by using a 16-bit UUID, and an almost infinite number[3] of proprietary services by using 128-bit UUIDs.

10.2.4.2 Units

Many of the values that are exposed represent physical values measured by a sensor. Therefore, it's useful to also define unit UUIDs for each of these possible types of value. The units are derived from the Bureau International des Poids et Mesures, otherwise known as the International System of Units (abbreviated SI from the original French, Système International d'Unités). This allows values captured from a Bluetooth low energy sensor to be used in other systems that also use the same SI units. It should be noted that even though the SI units are defined around the metric system, imperial units are also defined. So, even though velocity can be represented in meters per second, it can also be represented in kilometers per hour (km/h) or miles per hour (mph).

---

3. The number of services is approximately 10^{38}, more than enough for the planet earth for a few years.

10.2.4.3 Attribute Types

The most fundamental attribute types are allocated UUIDs from the Attribute Type UUID range. These are typically used for the attribute types defined by the Generic Attribute Profile, and not a service. The following attribute types are defined:

- Primary Service
- Secondary Service
- Include
- Characteristic

10.2.4.4 Characteristic Descriptor

Some data exposed by a service might include additional data. This additional data is labeled by using Characteristic Descriptors. An example of a descriptor would be a value that describes the format (the unit and representation) of an associated value.

10.2.4.5 Characteristic Types

This range of 16-bit UUIDs is the most used group of attribute types. Each unique type of value that is exposed by a service is allocated a Characteristic Type UUID. This allows a client to discover all the different types of data that a server has. Each characteristic type has a defined format and representation. There are a possible 22,015 characteristic types that can be defined, without having to resort to the almost unlimited number of 128-bit UUIDs that can also be used.

10.2.5 Databases, Servers, and Clients

A collection of attributes is called a *database*. A database can be very small and simple, the minimum being just six attributes,[4] or very large and complex. The complexity of the attribute database, however, is not at the attribute layer, it's how those attributes are used in services and profiles.

The database is always contained in an attribute server; an attribute client uses the Attribute Protocol to communicate with the attribute server. There is only ever one attribute server on each device, regardless of whether Bluetooth low energy or Bluetooth classic is used to make a connection with the other device. Because there

---

4. The smallest attribute database must have the following six attributes: ≪Primary Service≫ for ≪GAP Service≫, ≪Characteristic≫ for ≪Device Name≫, the ≪Device Name≫ value, ≪Characteristic≫ for ≪Appearance≫, the ≪Appearance≫ value, and a ≪Primary Service≫ for ≪GATT Service≫. This database doesn't expose much state and is therefore not particularly useful.

is only one attribute server on each device, there is only one attribute database on each device. For a Bluetooth low energy device, the attribute database includes a Generic Access Profile service that is mandatory to support. This means that every Bluetooth low energy device includes an attribute server and an attribute database (see Figure 10–11).

This means that the cost of exposing a small amount of information on a device—for instance, say just the battery state—is very small. This is because every device already includes an attribute database, so the only cost of adding in a service to expose this information is just the cost of three or more additional attributes. Given that each device starts with six attributes as a minimum, adding an extra three attributes for the battery service is fairly trivial.

Attribute Handle	Attribute Type	Attribute Value
0x0001	Primary Service	GAP Service
0x0002	Characteristic	Device Name
0x0003	Device Name	"Proximity Tag"
0x0004	Characteristic	Appearance
0x0005	Appearance	Tag
0x0006	Primary Service	GATT Service
0x0007	Primary Service	Tx Power Service
0x0008	Characteristic	Tx Power
0x0009	Tx Power	−4dBm
0x000A	Primary Service	Immediate Alert Service
0x000B	Characteristic	Alert Level
0x000C	Alert Level	
0x000D	Primary Service	Link Loss Service
0x000E	Characteristic	Alert Level
0x000F	Alert Level	"high"
0x0010	Primary Service	Battery Service
0x0011	Characteristic	Battery Level
0x0012	Battery Level	75%
0x0013	Characteristic Presentation Format	uint8, 0, percent
0x0014	Characteristic	Battery Level State
0x0015	Battery Level State	75%, discharging
0x0016	Client Characteristic Configuration	0x0001

Figure 10–11 An example of an attribute database

10.2.6 Attribute Permissions

Some attributes in an attribute server contain information that can only be read or written. To facilitate these restrictions upon access, each and every attribute in an attribute database also has permissions. Permissions themselves can be split into three basic types: access permissions, authentication permissions, and authorization permissions. Access permissions determine what types of requests can be performed on a particular attribute. Going back to our earlier examples, the state of the light might be readable and writable, the state of a phone call might only be readable, whereas the light control point might be writable only. Similarly, the state of a light might be readable to anybody but can only be written by trusted devices, the state of a phone call will require authorization to read its state, and the light control point will require authentication to write its state.

It should be noted now that attribute permissions only apply to the attribute value. They do not apply to the attribute handle or attribute type. Every device has permission to discover all the attributes that a device exposes, including their handles and their types. This is to allow devices to determine if a device supports something that it can use before authenticating and obtaining authorization. For example, it is possible to determine if a device supports the light control point attribute without authenticating. This makes the initial device and service discovery very user-friendly, while protecting the private and confidential information exposed by that device in those services.

The following access permissions are defined:

- Readable
- Writable
- Readable and Writable

When an attribute is read, the access permissions are checked to determine if the value of the attribute is readable. If it cannot be read, an error will be returned stating that the client cannot read this attribute value. Similarly, when an attribute is written, the access permissions are checked and if the value of the attribute cannot be written, an error stating that the client cannot write this attribute value will be returned.

The following authentication permissions are defined:

- Authentication required
- No authentication required

When an attribute is accessed, either for read or write, the authentication permissions are checked to determine if the attribute requires authentication. If it does require authentication, the client that sent the request is authenticated with this device. If the attribute does not require authentication, the value should be accessible, subject to other permission constraints. If the attribute does require authentication, only the clients that have previously authenticated will be allowed access. If a client is not authenticated with the device and it attempts to access an attribute that requires authentication, then an error stating that there is insufficient authentication will be returned.

If a client receives the insufficient authentication error, it can do one of two things: it can ignore the request and pass the error up to the application; or it can attempt to authenticate the client by using the SM and resend the request. It should be noted that the error code does not communicate the required level of authentication. Therefore, the client might need to either request authentication or raise the authentication level to gain access to the attribute value.

The interesting side effect of this behavior is that the client is in complete control over when and how authentication is performed. The server also doesn't need to hold the state of the received request. In Bluetooth classic, authentication is typically performed on the creation of an L2CAP channel. When the responder receives the channel request to a channel that requires authentication, it stores this request, sends back a pending response, initiates security procedures, and then finally resumes the original request. This is both complex and memory intensive. In Bluetooth low energy, the server simply responds as best it can to each and every request; the client contains the complexity of ensuring that authentication is performed, reissuing the original request again when necessary.

The following authorization permissions are defined:

- No authorization
- Authorization

Authorization is subtly different from authentication. It triggers similar behavior; an error response is sent with the error code *insufficient authorization* whenever an attribute access is attempted and the client is not authorized. However, this is an error that the client cannot resolve.

Authorization is a property of the server; the server either authorizes a client to access a set of attributes, or it does not. Therefore, it is up to the server to authorize clients. More important, the client has no signaling available to prompt the server to ask the user to authorize the client. Therefore, whenever a client attempts to access a given attribute that requires authorization, the server might prompt the user to authorize that client. The server might also immediately reject the request. The client would then need to wait before reattempting the request. Typically, the user of the client device will trigger the retry after he has configured the other devices to add the client to the list of authorized devices.

10.2.7 Accessing Attributes

Each attribute in an attribute database can be accessed by using one of the following five basic types of messages:

- Find Requests
- Read Request
- Write Request
- Write Command
- Notification
- Indication

Using Find Requests, a client can find attributes in an attribute database such that the more efficient handle-based requests can be used.

The Read Request is sent to read an attribute value. These either use one or more attribute handles or a range of attribute handles and an attribute type to determine which attribute value to read.

The Write Request is sent to write an attribute value. These always use an attribute handle and the value to write. It is also possible to prepare multiple values to be written before executing these writes in a single atomic operation.

Each of these requests always causes the attribute server to send a single response. If more data is required, another request must be sent by the client. For example, if the attribute value is very long and cannot fit into a single Read Response, the client can request the additional parts of the attribute value by using another Read Blob Request.

To minimize the complexity of the server, only one request can be sent at a time. Another request can only be sent after the previous response has been received.

It is also possible to use the Write Command to write an attribute value. This never causes a response. Because it does not have a response, this command can be sent at any time. This means that it is useful to write commands into a control point of an exposed state machine.

There are two additional types of messages, both of which are initiated by the server and send attribute values unprompted to the client. The Notification can be sent at any time and includes the attribute handle of the attribute that is being notified and the current attribute value of this attribute. The Indication is the same, having the same attribute handle and attribute value, but always causes an attribute confirmation to be sent back. These confirmations both acknowledge that the indicated value has been received but also that another Indication can be sent, whereas Notifications can be sent at any time.

10.2.8 Atomic Operations and Transactions

Each Attribute Protocol message that is sent from a client to a server, and vice versa, is sent as part of a single transaction. A transaction is either a single request followed by a single response or a single indication followed by a single confirmation. Transactions are important because they limit the amount of information that needs to be saved between successive transactions. This means that if a device receives a request, it doesn't need to save any information about that request to process the next request.

The other important aspect about the transaction model is that a new transaction can't be started until the last transaction has completed. For example, if a device sends a Read Request for an attribute, it can't send another until it has received the response from the last request. These transactions are only relative to a single device. A device that starts a transaction cannot initiate another transaction, but it can still process requests from peer devices.

There are a couple of exceptions to this simple rule: Commands and Notifications, and Prepare/Execute writes.

10.2.8.1 Commands and Notifications

There are two Attribute Protocol messages called Commands and Notifications with which a device can send a message to another device without having to await a response before sending another Command or Notification. These are useful when you must send a particular Command or Notification but are currently in the middle of another transaction. For example, suppose that you have have sent a Read Request to a particular device and are awaiting a response, and then you need to write a value on the same peer device. To do that, you would use a Write Command.

Commands and Notifications do not require a response or confirmation. This means that the sending device has no way of knowing if the message has been received and processed. For some applications, this is not acceptable, and a request/response or indication/confirmation is required. For some applications, however, this is perfectly acceptable. An interesting side effect of the lack of a response or confirmation is that the there is no limit to the number of these messages that a device can send. Effectively, a device can flood the peer device with Commands or Notifications. To protect against this, a device can drop any Command or Notification that it receives if it doesn't have the buffer space to store or process it. Therefore, these messages must be considered to be unreliable.

10.2.8.2 Prepare Write Requests and Execute Write Requests

The second exception to the preceding transaction rules is the Prepare Write Request and the Execute Write Request messages. Using these messages, a device can prepare a whole sequence of writes and then execute them as a single transaction. From the transaction point of view, each Prepare Write Request and response is a separate transaction. It is possible to interleave other requests in the middle of the complete sequence of prepares and execute.

There are two interesting side effects of this command: long writes and verification of writes. Each Prepare Write Request not only includes the handle of the attribute that will be written along with a value, but also the offset into that attribute's value where this part of the value will be written. This means that you can use a sequence of Prepare Write Requests to write a single, very large attribute for each part of the attribute value in a single execution.

The other interesting side effect is that the prepare write response includes the attribute handle, offset, and part value that was placed into the Prepare Write Request. This might at first appear to be a waste of bandwidth, but because values in the response will be the same as in the request, this protects against something going wrong.

Bluetooth low energy is sometimes a little protective of data; all bits in the payload are protected with a 24-bit cyclic redundancy check (CRC) that can detect up to 5 bit errors. If a packet is received that has 6 bit errors, there is a very small probability that the CRC will falsely accept this packet. The next guard is the 32-bit message integrity check (MIC) value that is included in every encrypted packet. This should reject a packet that has falsely passed the CRC value, but there is absolutely no guarantee that it will not also falsely pass an invalid packet. Therefore, there is an extremely small chance that a packet can be received that has falsely passed the checks.

Sometimes even a very small chance is too large. For example, if you are using Bluetooth low energy to control the sewage outflow valve of a city, you really don't want to write "close" only to discover that the valve received it as "fully open" and flooded the park and children's play area with... err, sewage.

It is for this reason that the prepare response includes the same data that was in the request. The fact that a packet is sent in two different directions, typically by using two different radio channels, each using different encryption packet counters, means that the chance that the same data in the response has been corrupted in the same way as the request is as close to zero as you could possibly make it. And, of course, if the response was wrong, then you can cancel the whole sequence of prepared writes by using the "cancel" code in the Execute Write Request and then start preparing to write again.

10.3 Grouping

The Generic Attribute Protocol only defines a flat structure of attributes. Each attribute has an address—its handle. However, modern data organization methodologies require significantly more structure than this simple flat structure. This is what the Attribute Profile enables. Instead of just a set of attributes, the Attribute Profile defines groups of attributes.

To understand why this is necessary, let's analyze how this could be done. It's possible to have "pages" of attributes. Each page would have a defined set of values. A page would be defined for each use case; for example, one page would describe the device, another page would be used if the device has a battery, and another page would expose the temperature. This is interesting if the devices are complicated. What happens when you have two batteries? What if there are two temperature sensors?

The biggest leap in software engineering over the last few decades has been the slow introduction of object-oriented paradigms. This essentially groups the data that describes an entity with the methods that you can use to control the data's behavior. The main benefit of using an object-oriented architecture is that each object is self-contained.

Let me take a moment to define some terminology. When talking about object-oriented programming, you might think of interfaces, classes, and objects. An interface is a description of external behavior. A class is an implementation of that interface. An object is an instantiation of that class. For example, a car is an instance of an automobile class that implements the driving interface, Not all car objects look the same; they can be implemented differently, but critically, they all have the same basic driving interface, such as the steering wheel, the accelerator pedal, and the brake pedal. The driving interface is the same, but the class that implements this interface can be different, and this can be instantiated many times, as is evident in traffic congestion.

Within Bluetooth low energy, grouping is used for both services and characteristics. A service is grouped by using a service declaration; a characteristic is grouped by using a characteristic declaration.

A service is a grouping of one or more characteristics; a characteristic is a grouping of one or more attributes.

10.4 Services

In software engineering, if you define and implement behavior for a given class, as long as the interface to that class is fixed, other parts of the system can reuse an object based on that class. This also means that if there is a bug in that class, you can fix it once, and all the other parts of the system can benefit immediately from that fix.

To ensure that classes are reusable, you must define an abstract interface that is immutable. Immutability is a strong word that means "unchanging over time." This immutability is the only thing that ensures the long-term viability of an interface. If interfaces were mutable, the user of that object would need to spend more time working out what interface that object has rather than actually doing what it needs to do.

Object-oriented systems typically use inheritance to enable changes to interfaces; a new class with a new interface inherits the behavior of an old class and then adds to or changes this behavior, as required. By ensuring that interfaces are immutable, they can be reused successfully for many years.

In Bluetooth low energy, the Generic Attribute Profile defines two basic forms of grouping: services and characteristics. A service is the equivalent of an object that has an immutable interface. Services typically include one or more characteristics, and can also reference other services. A characteristic is a unit of data or exposed behavior. These characteristics are self-describing, such that generic clients can read and display these characteristics.

Thus, a service is just a collection of characteristics and some behavior that is exposed through these characteristics (see Figure 10–12). The set of characteristics and their associated behavior encompasses the immutable interface for the service.

However, as Figure 10–13 illustrates, services can reference other services. And it is this simple concept that imbues enormous power to this architecture. A reference is just that—one service can point to another service. The reference can say many

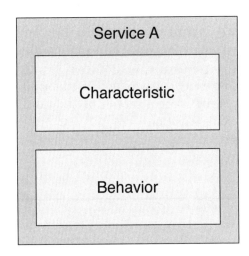

Figure 10–12 An immutable service interface is composed of characteristics and behaviors

10.4 Services

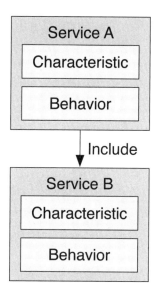

Figure 10–13 Service A references Service B

things: this service is used to extend the behavior of the original service; this service uses the other service; this service and the other service are combined together into a much bigger set of services. Let us examine each of these references in term.

10.4.1 Extending Services

Service A, which has been used for many years, now needs to be extended. Because Service A is immutable, we cannot simply add new behavior to the original service. Therefore, it's necessary to extend without altering the original service. To do that, we define a new service, Service AB, which contains the additional behavior required, as shown in Figure 10–14. However, to maintain backward compatibility for the many millions or billions of existing devices that only support Service A, we must also include an instance of that Service in every device that implements Service AB.

Now, suppose that we have two instances of Service AB, AB:1, and AB:2, on a device. We would also need two instances of Service A, A:1, and A:2 on the device. But which Service A belongs to which Service AB? To solve this problem, a reference needs to be made from each Service AB to the particular instance of Service A that it is explicitly extending (see Figure 10–15).

An old device that only understands Service A will still find the two old Service A instances and use them as before; the old device will ignore the other Service AB instances and, therefore, will only be able to use the nonextended behavior.

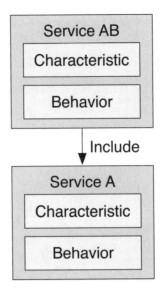

Figure 10–14 Service AB extends Service A

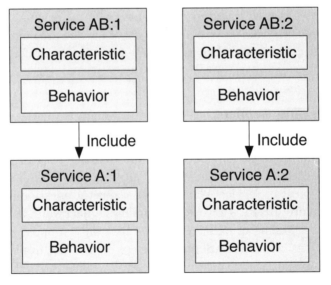

Figure 10–15 Two instances of Service AB extending Service A

A new device that understands both Service AB and Service A will find the Service AB instances and follow the references to their Service A instances. As such, the new device will be able to use the new behavior that is defined in Service AB.

10.4 Services

A new device that is talking with an old device will attempt to find Service AB, fail, and then find Service A; the new, therefore, will be able to automatically fall back to the interoperable behavior that was defined in Service A.

This appears complex, but it's actually much simpler than the alternative: Service A would be extended into a new version, with feature bits to determine which features a particular service supports and possibly very complex behavior because each possible combination of features would need to be tested. The extension methodology means that each service is self-contained and immutable, and the relationship between services is explicitly exposed. A device using the services can determine its behavior. Legacy compatibility is also guaranteed by the immutability of the original server.

10.4.2 Reusing Another Service

An additional method for reusing another service is to reference it. This is actually the simplest reference that can be made. One service, Service A, wants to use the behavior and state information from another service, Service B. To do this, Service A only needs to reference Service B. This is not reuse in the classic object-oriented sense; it is more like a generic pointer to another instance of a class.

This is useful because there may be many instances of both the referencing service as well as the referenced services: Service A:1, Service A:2, Service B:1, Service B:2, as depicted in Figure 10–16. Without the reference, it would be impossible to determine

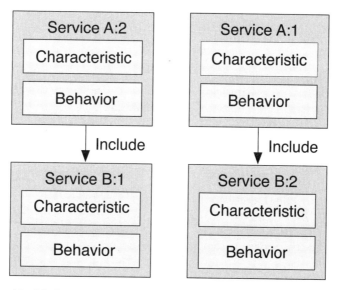

Figure 10–16 Service A reuses behavior and characteristics of Service B

if A:1 reused B:1 or B:2, or if A:2 reused B:1 or B:2. By including a reference to the other service, the particular instance of the service that is being reused will be known.

10.4.3 Combining Services

The final reference style is more complex than the other two in that it implies a separation of interface from implementation. Sometimes, it is necessary to have two independent service instances that are related to one another and have additional behavior when combined. To do this by using services, a third service must be defined that references both the original two services. For example, consider two instances of a service, Service A:1 and Service A:2, that need to be merged together and have additional "combinatorial" behavior. You can do this by instantiating a third service, Service C, that references both A1 and A2, as demonstrated in Figure 10–17.

Service C can expose the behavior that is required when dealing with both A:1 and A:2; it encapsulates the combined service behavior. For example, a light service and a daylight sensor service could be combined to give a service of a light that could only be switched on when there was no daylight.[5] The state machine of the combined service has extra states deriving from combinations of the state of the two basic services it references. The independent A:1 and A:2 services still have their

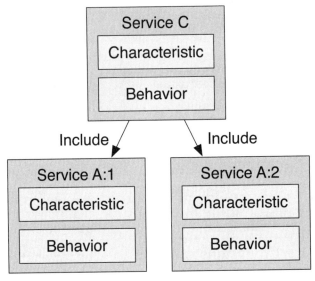

Figure 10–17 Service C combines the behavior of two instances of Service A

---

5. A more dramatic example would be an oven that automatically turned off when the food inside caught fire. Not that this happens in my house, of course :-).

own immutable behavior. This implies that Service C must very clearly distinguish the difference between the behavior associated with the combined services and the behavior of the independent services.

10.4.4 Primary or Secondary

One final concept to understand for services is that they can come in two different "flavors." Services can either be primary or secondary services. As you can clearly gather from the preceding description of the services and how they are designed, it is sometimes necessary to set up services that expose the external behavior of the device, and sometimes it is necessary to set up services that expose a block of functionality that can be reused many times in many different ways yet is never actually used or understood by the end user.

A service that exposes what a device does is typically a primary service. For example, if you have a device that supports Service B, Service B would be instantiated as a primary service. If you need some additional information for this device, which is available in Service D, but that information is not associated with what the device does, Service D would be instantiated as a secondary service (see Figure 10–18). So, a secondary service is an encapsulation of behavior and characteristics that are not something that a user would need to understand.

Primary services can be found quickly and efficiently by using the Attribute Protocol. They can have either a "parent" service or be a stand-alone service. Secondary services can only be found by reference and must always have another

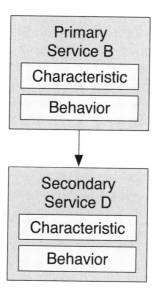

Figure 10–18 The relationship between primary and secondary services

206　　　　　　　　　　　　　　　　　　　　　　　　　　　　Chapter 10　Attributes

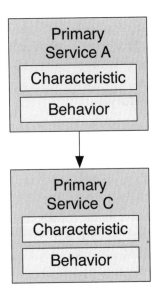

Figure 10–19 Primary services

service that points to them. This implies that a tree of services can be created, with a primary service at the top of each tree and each branch pointing to either primary or secondary services, and each branch from each of these being yet more primary or secondary services.

A primary service can point to another primary service, as illustrated in Figure 10–19. For example, the service extension would allow a new "version" of a service to be exposed and enable backward compatibility between these services.

A primary service can point to a secondary service so that it can reuse the behavior exposed in a secondary service. A secondary service can point to another secondary service or another primary service. Pointing from one secondary service to another secondary service is fairly rare because secondary services are typically leaf nodes in a service tree. Pointing from a secondary service to another primary service is extremely rare, but possible.

Primary services have one final advantage. When a client is looking for a particular service, it is possible to look for primary services very quickly. This can be further enhanced by only allowing a single instance of a given primary service on a device. For example, if a service is defined that can only have one instance of itself on a device, a quick search for that service by a client device would definitely determine whether that service exists.

This optimization has a significant benefit: A simple client that is only looking for one instance of a primary service can achieve that objective with the absolute minimum of fuss. Simple clients don't need to read the complete list of services in

10.4 Services

a device or determine their relationships to be able to use simple services. Without this optimization, every single simple client would need to walk the complete service tree to determine how it can use the services exposed on a device to best effect. This is a huge waste of valuable resources, both in terms of power for communication and memory to store all intermediate results and computations.

10.4.5 Plug-and-Play Client Applications

The other interesting aspect of the service model is that it is possible to take the set of service trees in a device and search for applications that can exploit these service trees. To do this, the generic client would begin by performing a complete service enumeration, first of the primary services and then following the relationships to other referenced services. Once this tree has been built up, it is possible to pass this "forest" of services to an application store to obtain the list of applications that are known to work with all, or part, of this forest.

Some applications might support just a single primary service. Some applications might support a primary service that extends another primary service, perhaps as an extension of the original service, and perhaps as a second version of the application. Some applications might support more than one service tree. These applications might be able to either present the information from these services in an interesting all-in-one application or combine this information together on the client in innovative ways.

For example, given a server that supports the services illustrated in Figures 10–18 and 10–19, it would support the primary Service A, including another primary Service C, the primary Service B including secondary Service D, and the primary Service C on its own: $A(C), B(D), C$. The client can then use this information to determine which applications support this set of services. The list of services for each App is then checked against this set to determine which applications can support this device. Some applications might only support a single service (App1 supporting C), whereas others might support the extended service A that includes C. Other apps might support both the $A(C)$ and B service trees, App6, whereas others might support the additional secondary service D included from B, App7 (see Figure 10–20).

Another approach is to use a generic application, App8, that can talk with any service. These client applications will typically not be able to interact as well as a specifically written application, but they might be able to support devices for which the client has no specific application already written.

This generic client behavior is explicitly supported by using the combination of services and a pure client-server model. However, the key element that makes this possible is the immutable services. Without services that have a known immutable behavior, generic client applications could not be written that can use this behavior.

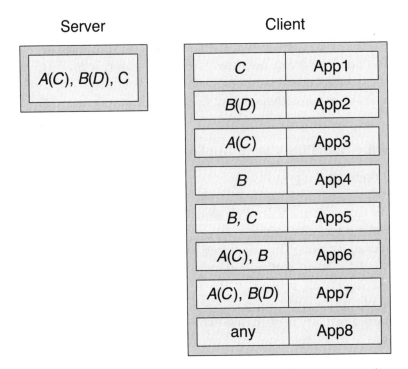

Figure 10-20 Services on a server mapped to applications on a client

The whole system has been designed for the maximum flexibility by limiting each individual part of the system to minimum flexibility. It is the combination of these individual immutable parts that provides the richness and ultimate flexibility required by products in the market.

10.4.6 Service Declaration

A service is grouped by using a service declaration (see Figure 10–21). This is an attribute with the attribute type of Primary Service or Secondary Service. All attributes that follow this service declaration and occur before the next service declaration are considered grouped with this service; they belong to this service.

As defined earlier, a primary service is one that encapsulates what the device does. A secondary service is one that helps the primary service achieve its behavior. All secondary services are referenced from a primary service. The reason for this is very simple; it retains simplicity of the client.

Simple clients are devices that have no user interface but can still use services on a peer device. A simple device can just search for the primary services and find the services that it needs. It does not need to walk the complete tree of services that a

10.4 Services

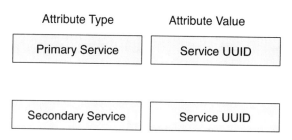

Figure 10–21 Primary and secondary service declaration

device might expose. In fact, the Attribute Protocol is optimized for simple clients by allowing them to search for a specific primary service.

Some services have helper services that assist them in exposing their behavior or state. For example, most medical devices will include device information; there is no need for each medical service to define their own device information. Similar device information is also required in the automation and battery scenarios. By defining this information in a service, the device information service only needs to be defined once and can then be used many times. This also makes it possible for those simple clients to not concern themselves with such information; they just ignore those secondary services when looking for their primary services.

The service declaration's value is a Service UUID. This is either a 16-bit Bluetooth UUID or a 128-bit UUID. Any service that a device does not understand can be safely ignored. For example, if a device includes a secondary service that has a 16-bit or 128-bit UUID that this device does not understand, all the attributes that are grouped with this service declaration can be ignored. To help with this, the Attribute Protocol allows the range of attribute handles for services to be discovered. Only known services will be processed further.

10.4.7 Including Services

Secondary services must be discovered separately. To do this, each service can have zero or more Include attributes. Include declarations always immediately follow the service declaration and come before any other attributes for the service. The Include definitions also encompass the handle range for the referenced service, along with the Service UUID for the included service (see Figure 10–22). This allows very quick discovery of the referenced services, their grouped attributes, and the type of the service. It does not state if this referenced service is a primary or a secondary service because this is not relevant.

Given that four octets are used for handles in the Include value, a Service UUID that is a full 128-bit UUID will not fit into the standard response packets used to find the included services. Therefore, when the included service has a 128-bit UUID, the

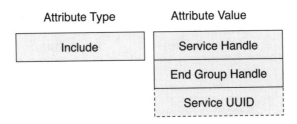

Figure 10-22 The structure of the Include declaration

Handle	Type	Value
0x0001	Primary Service	Service B
0x0002	Include	0x0200, 0x0209, Service D
...	...	...
0x0100	Primary Service	Service A
0x0102	Include	0x0300, 0x0317, Service C
...	...	...
0x0200	Secondary Service	Service D
...	...	...
0x0300	Primary Service	Service C
...	...	...

Figure 10-23 An example of an attribute database of Services A(C), B(D)

Service UUID is not a part of this declarations value. This means that an additional Attribute Protocol read is required to find the type of the service being included.

If the type of a referenced service—either primary or secondary—does not matter, a primary service can reference another primary service or a secondary service, and a secondary service can reference another secondary or primary service.

The preceding example that presented four services $A(C), B(D)$ would be created by using the database illustrated in Figure 10-23.

A primary service that was originally published that was later extended with another primary service would need to reference the original service. This original primary service cannot be changed to a secondary service because that would mean old clients would not be able to find that old service.

10.5 Characteristics

Grouping attributes together within a service demonstrates how these attributes can be combined to provide a consistent interface to a block of behavior. The architecture

10.5 Characteristics

of Bluetooth low energy also makes it possible to group attributes to allow the state and behavior of a service to be exposed.

A characteristic is just a single value. It could be current temperature, how far somebody has ridden their bicycle, or the state of the time synchronization finite state machine. However, a characteristic is much more than that. A characteristic needs to expose what type of data a value represents, whether a value can be read or written, how to configure the value to be indicated or notified or broadcast, and expose what a value means.

To do this, a characteristic is composed of three basic elements:

- Declaration
- Value
- Descriptor(s)

A declaration is the start of a characteristic; it groups all the other attributes for this characteristic. The value attribute contains the actual value for this characteristic. The descriptors hold additional information or configuration for this characteristic.

One question that is always asked at this point is why is the value an attribute within a characteristic and not just an attribute in its own right? The answer is actually fairly complex. A characteristic is not just a value; it has permissions, additional configuration, and descriptive data that is useful to consider as part of this characteristic. It could have been possible to add additional semantics to the Attribute Protocol to access this information, but this would have made the protocol more complex for the minority of cases for which this is really necessary.

Instead, a decision was made to keep the flat structure of attributes, as exposed by the protocol, separate from the structure of the device and its characteristics, as defined by the Generic Attribute Profile. This means that it is more complex to obtain certain information about some characteristics, but much easier to find the required information for most characteristics.

Simply put, a characteristic is composed of a characteristic declaration, the characteristic value, and zero or more descriptors.

10.5.1 Characteristic Declaration

To start a characteristic, a Characteristic attribute is used. This contains three fields: characteristic properties, the handle of the value attribute, and the type of the characteristic, as shown in Figure 10–24.

The characteristic properties determine if the characteristic value attribute can be read, written, notified, indicated, broadcast, commanded, or authenticated in a signed write. If the bit is set in this field, the associated procedure can be used to

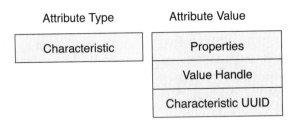

Figure 10-24 Characteristic Declaration

access the value of the characteristic value. Additionally, if the characteristic has the broadcast bit set, the server characteristic configuration descriptor must also exist. Similarly, if the characteristic has the notify or indicate bit set, the client characteristic configuration descriptor must exist.

There is also an extended properties bit in this field. This was added because there were additional properties to include in the 8-bit field, and the length of this field as the maximum size of this Descriptors value was already met. These additional properties are in the characteristic extended properties descriptor. There are only two additional properties: reliable write support and writable auxiliaries. The writable auxiliaries is the most interesting bit because this determines whether the characteristic user description descriptor can be written.

The characteristic value handle field is the handle of the attribute that contains the value for the characteristic. This is needed so that a very quick search for the characteristic can be performed by a client that returns only characteristic declarations. With this declaration, the attribute that holds the value is immediately available. If this field did not exist, the client would then need to perform an additional search for attributes and effectively guess which attribute after the declaration was the value. At the moment, the value attribute is the very next attribute after the characteristic declaration, but by including the handle for the value attribute in the declaration, this practice could be changed in the future.

The final field is the characteristic UUID. This holds the UUID that is used to identify the type of the characteristic value. This UUID must be the same as the type of the attribute that holds the characteristic value. Effectively, this means that this is a duplication of information that could be determined by sending more requests to the server. However, this would require more over-the-air protocol messages to be sent, wasting power. It is more efficient to include the type information in the declaration directly.

On occasion, it has been questioned why the characteristic value attribute's type is not a static UUID such as Value. This would reduce the previously described

10.5 Characteristics

Handle	Type	Value
0x0001	Primary Service	GAP Service
0x0002	Characteristic	read write, 0x0003, Device Name
0x0003	Device Name	"Proximity Tag"
0x0004	Characteristic	read, 0x0005, Appearance
0x0005	Appearance	Tag

Figure 10–25 Characteristic example

problem; however, there are other optimizations that can be performed, indicating that this would not be an ideal solution (see Figure 10–25). For a simple client that only wants to retrieve the battery state of a device, it would be much more efficient to just ask for the battery state rather than search for the characteristic that has the battery state UUID in one of its fields. It is these simple optimizations that have determined the structure of the declaration, as demonstrated in the following:

```
// The complicated way
service_range = discover_primary_service_by_UUID («Battery_Service»)
chars = discover_all_characteristics_of_a_service (service_range)
foreach char in chars:
   if char.uuid == «Battery Level»:
       battery_level = read_characteristic_value (char.handle)

// The easy way
battery_level = read_characteristic_value_by_UUID («Battery Level»)
```

10.5.2 Characteristic Value

The characteristic value is an attribute with the type that must match the characteristic declaration's characteristic UUID field. Apart from that, it is an ordinary attribute. The biggest difference is that the types of actions that can be performed on this characteristic value attribute are exposed in the characteristic declarations properties field and additionally might be in the characteristic extended properties descriptor.

For each characteristic, a specification document can be found that describes the format of the characteristic. Also, characteristics themselves have no behavior, so the service specification with which this characteristic is grouped should be examined to determine the behavior exposed by this instance of the characteristic.

10.5.3 Descriptors

There can be any number of descriptors on a characteristic. Most descriptors are optional, although, as just explained, they might be required depending on the characteristic declaration. Some descriptors might also be required by a service specification.

The following descriptors can be included in a characteristic:

- Characteristic Extended Properties
- Characteristic User Description
- Client Characteristic Configuration
- Server Characteristic Configuration
- Characteristic Presentation Format
- Characteristic Aggregation Format

10.5.3.1 The Characteristic Extended Properties Descriptor

This is the descriptor that is used to capture the additional extended properties. At the moment, only two are detailed: the ability to perform reliable writes on the value and the ability to write the Characteristic User Description descriptor.

10.5.3.2 The Characteristic User Description Descriptor

Using this descriptor, a device can associate a text string with a characteristic. This is most useful with devices for which users can perform this configuration themselves. For example, the user could configure a thermostat to describe which room in the building the device is measuring. Some devices might include multiple temperature sensors, so having this configuration at the characteristic level is essential for the ultimate configurability.

10.5.3.3 The Client Characteristic Configuration Descriptor

If a characteristic is notifiable or indicatable, this descriptor must exist. It is a two-bit value, with one bit for notifications and the other for indications. Notification and Indication are complementary procedures, so it is impossible to set both of these bits at the same time. How the value is notified or indicated is not defined in the core specifications; this is defined by the service specifications.

10.5.3.4 The Server Characteristic Configuration Descriptor

This descriptor is very similar to the Client Characteristic Configuration descriptor, except that it has one bit for broadcast. This is a single bit, and setting it causes the device to broadcast some data associated with the service in which this characteristic is grouped. Again, the timing of this broadcast is determined by the service.

Interestingly, it is not possible to broadcast a single characteristic. Instead, the service for which this characteristic is grouped defines what data is broadcast when this bit is set. Some services might define that multiple characteristics can be broadcast; it is up to the service to define how an observer can determine which characteristics are broadcast by the service.

It might appear at first to be rather strange that there is bit in a characteristic that can turn on the broadcast of this characteristic, without having the ability to actually broadcast the characteristic directly. This is because characteristics themselves do not have behavior; thus, the meaning of broadcast characteristic data without the context of a service is meaningless. Just receiving "Temperature : 20.5°C" doesn't mean much. Receiving "Room Temperature Service : 20.5°C" or "Car Engine Service : 65°C" gives that temperature the needed context.

10.5.3.5 The Characteristic Presentation Format Descriptor

One of the goals for the Generic Attribute Profile was to enable generic clients. A generic client is defined as a device that can read the values of a characteristic and display them to the user without understanding what they mean. A generic client could connect to a refrigerator and display the inside temperature without understanding that a value above 10°C is probably bad. In contrast, a profile defines how a client can interoperate with a temperature service in a refrigerator and what to do when the temperature goes out of a valid range.

For generic clients to work, they must be able to find characteristics that can be displayed to the user and then understand their characteristic values enough to display them to the user. The characteristic declaration having a known attribute type is one aspect of being able to find all the characteristics within a device. Generally, characteristics that are readable are also useful. The most important aspect that denotes if a characteristic can be used by a generic client is the Characteristic Presentation Format descriptor. If this exists, it's possible for the generic client to display its value, and it is safe to read this value.

The presentation format is a multiple-field value that contains the following fields:

- Format
- Exponent
- Unit
- Namespace
- Description

The format is an enumeration of the standard data types that determine how the value is structured. There are formats for Boolean and unsigned 2-bit and 4-bit formats. There are formats for both unsigned and signed integer values with sizes ranging from 8 to 128 bits. There are two sized standard IEEE-754 floating-point numbers, such as are used in most high-end computers. There are two sized integer-based fixed-point numbers that are used primarily by medical devices. Finally, there are two string representations using both UTF-8 and UTF-16 encodings. If the format of the characteristic doesn't fit into one of these buckets, the opaque structure can be used, or an aggregate format should be used, as defined in the following description.

After the format comes the exponent. This field is only valid for the integer values; it determines a fixed exponent that can be applied to the integer value before it is rendered to the user. This is a base 10 exponent, which makes it possible to perform the placement of the decimal point in the final output routine, rather than using complex mathematics. The value that the characteristic value represents can be expressed by using the following formula:

$$displayed\ value = characteristic\ value * 10^{exponent}$$

For example, if the characteristic value is 0xFD94, and the presentation format is a signed 16-bit integer with an exponent of -2, the displayed value will be as follows:

$$-620 * 10^{-2} = -620 * 0.01 = -6.20$$

The next field in the presentation format is the unit field. The unit is a UUID defined in the assigned numbers document. Many units are defined. For example, in the preceding example, if the unit is Temperature Celsius, the displayed value will be $-6.2°C$. It is obviously assumed that a generic client knows what each of these unit UUIDs are.

The final two fields should be considered as a single value. The namespace and description fields determine additional information about the value. The namespace field is a single byte that determines which organization controls the description field. The description field is a 16-bit unsigned number.

The description field is really just a single "adjective" that can be applied to the characteristic so that the user can determine which value is associated with a particular property of the device. As an example, consider a thermometer that has both inside and outside temperature probes. This would expose two temperature characteristics, the only difference being the description field of the Characteristic Presentation Format descriptor being either "inside" or "outside."

The description and unit fields are used as a lookup to a string localized to the user's language. Therefore, for a characteristic that has the unit of "weight (kg)" and a description of "hanging", this localized string would be one of "hanging weight", "hengende vekt", "vješanje težina", "riippuva paino", "penjant de pes", depending on the user's language.

10.5.3.6 The Characteristic Aggregation Format Descriptor

Some characteristic values are more complex than just a single value. For example, look at the standard denotation of a position on the planet earth. This is composed of two values concatenated together into a single "value." The position value is an aggregation of a latitude value and a longitude value. To allow for such complex characteristic values, the Characteristic Aggregate Format descriptor allows multiple presentation format descriptors to be referenced so that the individual fields of the value can be illustrated.

Using the preceding example, the characteristic would have two Characteristic Presentation Format descriptors (one for the latitude and one for the longitude) and the Characteristic Aggregation Format descriptor that references these two Characteristic Presentation Formation descriptors in their correct order. A generic client can then correctly deconstruct the format of the characteristic value and display the value to the user.

It should be noted that there is no requirement for these Characteristic Presentation Format descriptors that are referenced from the Characteristic Aggregate Format to be in the same characteristic. They might not even be in the same service or device. They are just referenced presentation formats; the characteristic within which they are grouped has no meaning for the aggregate format.

10.6 The Attribute Protocol

The Attribute Protocol is a very simple protocol by which an attribute client can find and access attributes on an attribute server. It is structured as six basic operations:

- Request
- Response
- Command
- Indication
- Confirmation
- Notification

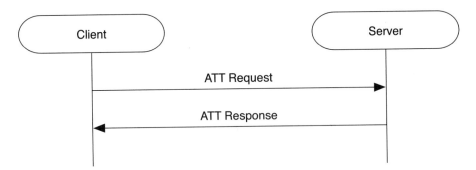

Figure 10–26 An Attribute Protocol request

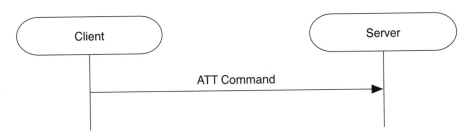

Figure 10–27 An Attribute Protocol command

The client sends a request to the server to request that the server do something and send back a response, as depicted in Figure 10–26. A client can send only one request at a time. This reduces the complexity on the server, reducing the memory requirements, and thereby making it possible to implement an attribute server using very little code. For each request, there can be only two possible responses: a response that is directly associated with the request, or an error response that gives information about why the request failed.

A client also sends a command to a server but receives no response, as demonstrated in Figure 10–27. The client uses commands when it wants the server to perform an action but there is no need for a immediate response, for example, when the client commands the server to change the television channel. A command can also be used when the response might be delayed and therefore would be delivered in the form of indications or notifications.

Indications are sent by a server to a client to inform the client that a particular attribute has a given value (see Figure 10–28). Indications are similar to requests in that a confirmation response is required by the client. Also, the server can send only one indication at a time, meaning that only after receiving a confirmation for a previous indication can the next indication be sent.

10.6 The Attribute Protocol

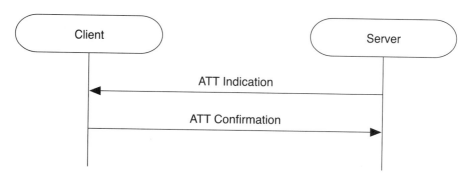

Figure 10–28 An Attribute Protocol indication

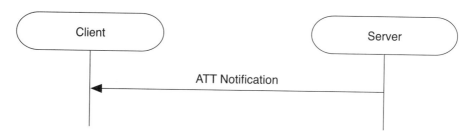

Figure 10–29 An Attribute Protocol notification

The server sends notifications to a client to inform the client that a value of a particular attribute has a given value (see Figure 10–29). Notifications don't require a response. In this way, they are similar to commands.

Because commands and notifications have no response or confirmation, there are no restrictions on how often they can be sent. If too many commands or notifications are sent, such that the server or client cannot process them all, the receiver of the messages can discard them. They are therefore unreliable. Requests and indications are therefore considered to be reliable because they all must elicit a response acknowledging that the receiving side has at least processed the request or indication.

10.6.1 Protocol Messages

Figure 10–30 provides a list of all the Attribute Protocol PDUs. For most messages, there are both a Request and a Response PDU. For example, the Read message has a Read Request PDU and Read Response PDU. For each of these, there is a set of parameters, which are summarized here.

The following sections refer to the example attribute database that is shown in Figure 10–11.

Message	Request Parameters	Response Parameters
Error		Opcode in Error, Handle in Error, Error Code
Exchange MTU	Client Rx MTU	Server Rx MTU
Find Information	Starting Handle, Ending Handle	(Handle, Type)
Find By Type Value	Starting Handle, Ending Handle, Type, Value	(Found Handle, End Group Handle)
Read By Type	Starting Handle, Ending Handle, Type	Length, (Handle, Value)
Read	Handle	Value
Read Blob	Handle, Offset	Part Value
Read Multiple	(Handle)*	(Value)
Read By Group Type	Starting Handle, Ending Handle, Group Type	(Handle, End Group Handle, Value)
Write	Handle, Value	-
Prepare Write	Handle, Value	Handle, Value
Execute Write	Flags	-

PDU	Command Parameters
Write Command	Handle, Value
Signed Write Command	Handle, Value, Authentication Signature

PDU	Notification Parameters
Handle Value Notification	Handle, Value

PDU	Indication Parameters	Confirmation Parameters
Handle Value Indication	Handle, Value	

Figure 10–30 Attribute PDUs

10.6.2 The Exchange MTU Request

The Attribute Protocol has a default maximum transmission unit (MTU) of 23 octets on a Bluetooth low energy link. If a device wants to send larger packets, it must negotiate a higher MTU size. Only the client can initiate this request. However, given that many devices have both client and server functionality, this should not be a critical issue. The client request includes the client receiver's MTU size; the server request includes the server receiver's MTU size. These two values cannot not be the same value. The MTU size that a link uses can be calculated by taking the minimum of both the client Rx MTU and the server Rx MTU.

This value is not negotiated. In fact, the values sent in the Exchange MTU Request and Exchange MTU Response are fixed values so that the implementation is made as easy as possible. A device that is both a client and a server must use the same value for its client Rx MTU and server Rx MTU, meaning that regardless of which device initiates the MTU exchange, the same MTU results. This means that in the event that both devices initiate the procedure at the same time, the result is the same as if one or the other had started it at different times. This restriction also means that there is no point in starting another MTU exchange at a later time during a connection because the result will always be the same.

For very simple devices that don't support anything more than the default MTU, the server can always respond with a service Rx MTU of 23, and nothing changes. Obviously, a client would not initiate an MTU exchange if it only supports the default MTU. This implies that all devices must support at least the default MTU.

10.6.3 The Find Information Request

The Find Information Request and response are used to find handle and type information for a sequence of attributes (see Figure 10–31). This is the only message that enables a client to discover the types of any attributes.

The Find Information Request includes two handles: a starting handle and an ending handle. These define the range of attribute handles used for this request. To find all value attributes, the request would have the starting handle set to 0x0001 and the ending handle set to 0xFFFF. Typically, the response can only include a few attributes within this range of handles at a time; therefore, a sequence of these Find Information Requests must be performed to find all the attributes, with the starting handle one higher than the last found attribute.

The Find Information Response contains handle-type pairs. There are two possible formats for this because there are two sizes of UUID used in Bluetooth low energy. One format is for 16-bit UUIDs that allows up to 5 attribute handle-type pairs to be included in a single Find Information Response. The other format is for 128-bit UUIDs that can only contain a single handle-type pair in the response.

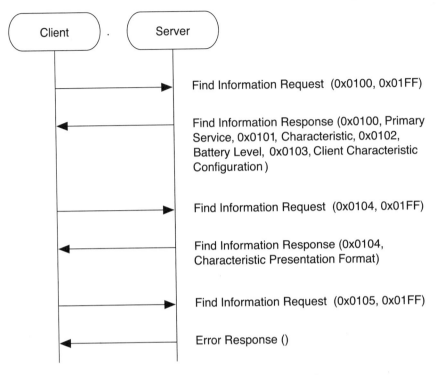

Figure 10–31 The Find Information Request

Obviously, with a larger MTU than the default, the number of handle-type pairs that can be included increases.

The other interesting thing about the response is that it cannot include both 16-bit and 128-bit UUIDs in the same response packet. To do this would have required an extra byte for each handle-type pair, reducing the number of 16-bit UUIDs that could have been included in a default MTU response to just 4. Given that most UUIDs in a standard attribute server will be 16-bit UUIDs, and it is also not possible to include both a 16-bit UUID and a 128-bit UUID in a single response, it was not considered useful to provide this flexibility.

10.6.4 The Find By Type Value Request

The Find By Type Value Request and response can find all the attributes with a given type and value. The request includes two handles: a starting handle and an ending handle. These define the range of attribute handles used for this request. Any attribute with a handle of starting handle up to and including ending handle that has the same type and same value as the request is returned in the response.

10.6 The Attribute Protocol

The response includes one or more handles for each attribute that is found. If the type in the request is considered a grouping attribute, one of Primary Service, Secondary Service, or Characteristic, the handle of the last attribute within this group is also included in the response. Unfortunately, because the format of the characteristic declaration does not have a static value for a given characteristic, it is not possible to use this request to search for characteristics.

The primary use of this request is to find a specific primary service. A client can send a Find By Type Value Request with the type set to Primary Service and the value set to the UUID of the service. The response then includes the handle range of each instance of this primary service that is found. Some services are specified such that they cannot exist in a server more than once; therefore, this request would only return a single handle range for that service.

It is possible to use this request to find secondary services; however, this is not something that is used by Bluetooth low energy today. Secondary services are always included from other services, so the Read By Type Request is used to discover these services.

10.6.5 The Read By Type Request

The Read By Type Request reads the value of an attribute within a range of handles. The client uses this when it knows only the type, not the handle. The request includes a starting handle and ending handle. It also includes the type of the attribute to be read. The response includes pairs of handles and values.

Each attribute within the handle range that has the requested type is returned. The response is a set of attribute handles and their associated values. The response is optimized for attributes that have the same size values. For example, when reading all the includes of a service, the Include attributes might all have the same sized values; thus, a single response is returned. If the attributes have different sized values, only the first one or first few that have the same length of value will be returned in the first response. If the attributes have the same sized values, but only some fit into a single response, only the ones that fit into the response will be returned. The request will then have to be repeated with an updated starting handle to obtain further requests.

The Read By Type Request is used for searching for included services as well as discovering all the characteristics of a service by using the Characteristic type. It is also used to read the value of a characteristic with a known type. For example, if you just want to quickly read the battery level of a device, a quick method is to use a read by type request, with the type set to Battery Level. The response will then include the handles of the characteristic values for the battery state along with its values. This allows a client to read out some data very quickly by sending a quick request to it and get that attribute type's value without extensive characteristic discovery.

10.6.6 The Read Request

The Read Request is probably the simplest request in the Attribute Protocol. The request includes a handle, and the response returns the value of the attribute identified by that handle. This is only useful when the attribute handle is known, but assuming that this knowledge is available on the client, the value of that attribute is read.

10.6.7 The Read Blob Request

Sometimes, the value of an attribute is longer than can be contained within a Read Response. In this situation, the Read Blob Request can be used to read the additional bytes of an attribute's value. The word *blob* comes from the database term meaning Binary Long Object. The Read Blob Request includes not only the attribute handle but also a zero-based offset into the attribute value. The response contains as much of the value of the attribute as from the given attribute offset.

This request can be used after the Read Response has returned the first 22 octets of an attribute when the client expects a longer attribute value. That Read Blob Request uses the same attribute handle but with an offset of 22. The response includes part of the value of this attribute from offset 22 to the byte at offset 43. This can continue until the complete attribute value has been read.

This is used when reading long characteristic values and reading long characteristic descriptors.

10.6.8 The Read Multiple Request

The Read Multiple Request reads multiple attribute values in a single operation. The request includes a set of one or more attribute handles. The response includes the values of these attribute handles in the order that they were requested. This is optimized to read multiple attributes of a known size. For example, weight scale might measure an individual's weight as well as a body mass index; both of these values can be read in a single request.

Because there is no delimitation between the values in the response, variable-length attribute values cannot be used, with a single exception. The last attribute requested can have a variable length. This means that if a client requests a read of three attributes in a single-read multiple request, the first two attributes must have a fixed size, whereas the last attribute can have a variable length.

If a client requests multiple attributes whose values would extend beyond the end of the response packet, the values that don't fit into the response will be silently dropped.

10.6.9 The Read By Group Type Request

The last read request is the Read By Group Type Request. This is very similar to the Read By Type Request in that it takes a range of handles that the read will be considered over as well as an attribute type. The difference is that the attribute type must be a grouping attribute type, and the response includes the handle of the read attribute, the last attribute for that grouping attribute, as well as the value.

This means that if the grouping type is a Primary Service, it will return not only all the primary service declaration attribute handles but also the last attribute within that primary service and the value of the primary service declaration. A single request can therefore be used to discover all the primary services within a device, and the handle ranges of all the attributes associated with those services, and the type of those services.

As with the other responses where pairs of handles and values are returned, if the values have a variable length, only the first attributes with the same length will be returned in a single response. Therefore, it is necessary for the client to send the request again, updating the starting handle to find the next attribute of interest.

10.6.10 The Write Request

The Write Request is analogous to the Read Request. The request includes a handle and the value to be written into that attribute. The response acknowledges that the value was written.

10.6.11 The Write Command

The Write Command is similar to the Write Request, except there is no response sent. The Write Command contains the handle of the attribute to be written along with the value that is to be written.

Use this command when there is no need for a response. Also, because this command can be sent at any time, even after another request has already been sent and an associated response has not yet been received, this command is also useful in situations for which the latency of sending the command is very important.

10.6.12 The Signed Write Command

The Signed Write Command is very similar to the Write Command except that it also includes an authentication signature. This way, the sender can authenticate itself along with the handle and value being commanded to the server without the need to encrypt the link. This is most useful when the cost of starting encryption would either significantly increase the latency of the data connection or increase the cost of sending a small bit of data when the data does not have to be confidentially delivered.

The authentication signature is composed of a SignCounter and a message authentication code. The SignCounter must be a different value for each message sent between the two devices, regardless of whether the link was disconnected between messages being sent. The message authentication code is the result of the CMAC function as defined in the NIST special publication 800-38B, using a Connection Signature Resolving Key (CSRK) that can be distributed when two devices bond.

The SignCounter is a 32-bit value, which means that there are 4 billion possible signed writes to be performed before the devices need to distribute a new CSRK. Because the SignCounter protects against replay attacks, a Signed Write Command that is received with the same SignCounter must be ignored. The message authentication code is a 64-bit value that is appended after the handle, value, and SignCounter. It should be noted that this requires that the server stores the last SignCounter for each client.

10.6.13 The Prepare Write Request and Execute Write Request

The Prepare Write Request and Execute Write Request are used for two purposes. First, they provide the ability to write long attribute values. Second, they allow multiple values to be written as a single-executed atomic operation.

The attribute server contains a single prepare write queue in which Prepare Write Requests are stored. The size of the queue is implementation dependent, but typically, it is large enough for all the expected services that require prepared writes. The values that are prepared are not written into the attributes until an Execute Write Request is received that gives the go-ahead to execute these prepared writes.

The Prepare Write Request includes the handle, offset, and part attribute value in a similar way as that of the Read Blob Request. This means that the client can either prepare the values of several attributes in the queue or prepare all the parts of an attribute value to be written in the queue. This way, a client can be sure that all the parts of an attribute can be written on the server before the prepare queue is actually executed.

The Prepare Write Response also includes the handle, offset, and part attribute value from the request. This is pure paranoia to ensure that the data gets through reliably. The client can check the fields in the response, with the values it placed into the request, to ensure that the prepared data is received correctly. For some applications, this is certainly a level of verification that is needed. As per the earlier sewage valve example, it is better to send a message twice—and check it—than to flood the children's play park with untreated effluent because of a single bit error.

To see how careful a client can be about writing a single byte of data to a server, we need to examine how much protection there can be on this data. When a client sends a Prepare Write Request to a server on an encrypted link, there is a 24-bit CRC making sure that there are no bit errors as well as a 32-bit MIC value to ensure that it was sent from the correct client. The prepare write response is also protected with the same 24-bit CRC and 32-bit MIC values.

This means that a single Prepare Write Request of a 1-byte attribute value is being protected by 112 bits of error-checking values: a ratio of 14 protection bytes to the single data byte. In addition, if the client is not happy with the Prepare Write Response, it can just send a quick Execute Write Request with the flags set to "cancel" the Prepare Write Queue and start preparing the value again.

Once all the prepared writes have been sent, the server has a queue of prepared writes ready to execute. The client can then send the Execute Write Request with the flags set to immediately write all prepared values. The server will then write all these values in a single atomic operation. The attributes are written in the order in which they were prepared. If the client prepared the same attribute multiple times, the server will write these values in order. This means that if the prepare queue is used to configure a hardware state for which the hardware must be disabled, configured, and then reenabled, this can all be done with in a single prepare queue by writing "disable" to the appropriate attribute, writing the configuration attributes, and then writing "enable" at the end of the prepare queue. The execution of this prepared queue will therefore complete the reconfiguration in an atomic operation.

10.6.14 The Handle Value Notification

When the server wants to send a quick attribute state update to the client, it can send a Handle Value Notification. This is one of only two messages that the server can send to the client, and the only one that doesn't have a message sent in reply. Therefore, the server can send it at any time and it is also considered to be unreliable.

The Handle Value Notification has an attribute handle and a value. The notification is therefore a message from the server to the client that this attribute now has this value. This message is one of the most important messages in the Attribute Protocol. It not only allows the server to efficiently keep the client up to date with the current state of its attribute database, but it is also used to notify the client of the changes in a finite state machine.

Without notifications, a client would need to constantly poll the server to determine if the attribute value had changed. Once notifications are configured, the client just waits for the server to notify it when the value has updated. It should be noted that this also means that the client is be notified of the new value more quickly than if it were polling periodically.

Typically, the configuration of notifications within a server is stored for all bonded devices. This has the advantage that when a client reconnects to a device, the server in that device can instantly notify the client of its state. For example, the battery level of a device could be configured for notification; whenever the client reconnects to the device and the battery level has changed, the client is instantly notified.

Because these notifications don't have any confirmations, they can be sent at any time, regardless of any other transactions that are active at the same time. This means that even if the client and server are in the middle of a complex interaction involving requests and indications, a notification can always be sent. As such, notifications are very useful for information that needs to be sent to the client now.

10.6.15 The Handle Value Indication

A Handle Value Indication is very similar to the Handle Value Notification; it has the same fields of attribute handle and value, but it must be confirmed upon receipt in the client. The server can only send one of these indications at a time, and it will only send the next indication when the last indication has been confirmed.

The handle value confirmation doesn't have any data within it. It is used for flow control. Because of the confirmation, the indications can be considered to be reliable. Once the server has received the confirmation, the server can be assured that the client has received the indication.

10.6.16 Error Response

An Error Response can be sent by a device whenever a request asks for something that cannot be achieved. For example, if a device asks for an attribute to be written by using a write request, but that attribute is read-only, instead of sending a write response saying that everything was good, an Error Response is sent giving the reason the request failed.

The Error Response includes all the information about the request that caused the error, the attribute on which the request failed, and why the error was generated in the first place. Whenever a client receives an error response, it must assume that this error response is for the last request that it sent. Therefore, the Error Response is another way to close the request's transaction along with its own response message. This means that for each request that can be sent, there are always two possible responses, the failure error response case and the success response case. For example, a Read Request can either have an Error Response or a Read Response sent in reply.

The following are the different reasons an error can be sent:

- **Invalid Handle** The attribute handle in the request was invalid and does not exist on the server. This could be because a read or write was requested using an attribute handle that is not used or allocated by that server. It can also be sent when an attribute handle in the request was set to 0x0000.

- **Read Not Permitted** The attribute does not allow the reading of the attribute value. For example, a control point write-only attribute will send this error if a client attempted to read this value.

- **Write Not Permitted** The attribute does not allow the writing of the attribute value. For example, attempting to write a read-only attribute value will send this error.

- **Invalid PDU** The request that is sent is not understood by the server. This is typically sent when the client has sent a request that is badly formatted. For example, a Read Request should have a two-octet handle as its single parameter. If the Read Request does not have two octets of parameters, this error is sent.

- **Insufficient Authentication** The request to read or write an attribute's value cannot be completed because the two devices have not authenticated one another. To perform authentication, the connection can be encrypted, or if no bond exists for that device, the Security Manager pairing procedures can be used to pair and then bond the devices.

- **Request Not Supported** The request that is sent is known to the server but it has chosen not to implement it at this time. This can be used for both known requests that the server did not implement as well as future requests that it does not currently understand. This error can be the default error on any request that is not implemented in a device.

- **Invalid Offset** The request includes an offset, but the offset given was invalid. When reading long attributes, the offset is used to read the parts of the attribute value one block of bytes at a time. If the offset used is greater than the length of the attribute value, the offset would be considered invalid. An offset the same as the length of the attribute value does not give this error; instead, it would respond with zero-length part value in the response.

- **Insufficient Authorization** The request to read or write an attribute's value cannot be completed because the server has not authorized the client to have access to that value. Authorization is a server local feature. The server needs to determine how to allow the user to authorize that client. For example, on a phone or a computer, this could be done by prompting the user and asking her if she grants access to the requested data.

- **Prepare Queue Full** The Prepare Write Request cannot be accepted because the memory used to hold the queue of pending writes is already full. The prepare queue is a finite size and should be sufficiently large to handle all the services that the device supports. However, a client might attempt to do too much in a single prepared queue and therefore would fail.

- **Attribute Not Found** The sought-after attribute was not found. This is only used when searching over a range of attributes when looking for a specific attribute type or types. This means that only the Find Information Request, Find By Type Value Request, Read By Type Request, and Read By Group Type Request can generate this error. For the Find Information Request, this means that there were no attributes found within the handle range. For the Find By Type Value Request, Read By Type Request, and Read By Group Type Request, this means that there were no attributes of the given type found within the handle range.

- **Attribute Not Long** The attribute referenced in the Read Blob Request is not a long attribute and therefore the request is rejected and the client should use the Read Request instead. This error is only relevant to fixed-length attribute values, which are shorter than the current MTU size where the Read Blob Request is used. For this reason, it is much simpler to use a Read Request to read the first 22 octets of an attribute value and then use the Read Blob Request to read the remaining octets.

- **Insufficient Encryption Key Size** This error is generated when the link is encrypted and has sufficient authentication and authorization, but the encryption key size negotiated during pairing is weaker than that required by the service. There are some attribute values where very strong encryption keys are necessary to fully protect the confidentiality of the data.

- **Invalid Attribute Value Length** The attribute value used in the write request or from the prepare queue during the execution of that prepare queue is the wrong length. If an attribute value has a fixed size (for example, two bytes) and the write attempts to change this to a different size (for example, one or three bytes), then this error is generated.

- **Unlikely Error** This is probably the best error code that exists. It basically means that something unexplained happened that doesn't fit any other error code. The problem with this error code is that if the error was describable, an error code could be created for it. Given that the error generating this was not thought of ahead of time, it really is an unlikely error.

- **Insufficient Encryption** The attribute value can be read or written but the link is not encrypted. There are some attribute values for which the confidentiality of an encrypted link is required.

- **Unsupported Group Type** The attribute type that was included in the request was not considered a group type by the server. Only group types that are known by the server can be used in Read By Group Type Request.

- **Insufficient Resources** The server has insufficient resources to accept or process this particular request. For example, some services might use this error code when configuring a device to broadcast data when the quantity of data that is already being broadcast in addition to the requested broadcast data is too large.

- **Application Errors** The request did something with an attribute of a service that was not allowed, and the service allocated its own error code to report what went wrong. This is a range of error codes, from 128 to 255, and the actual meaning is defined in the service specification in which the attribute is grouped. A typical application error would be that the value written into a characteristic was invalid.

It should be noted that an error response terminates the request. If the client fixes the error, perhaps by authorizing or performing a procedure that authenticates the link, the client will have to repeat the request. There is no "pending" response. Pending responses assume that the server can hold state about the request while it sorts out the problem. In Bluetooth low energy, the client is always the more complex device so it must resolve the problems and hold the state again. It also means that a client can send any request to the server, knowing that it can always receive an error, and then move on to the next attribute if it is unable to fix the problem causing the error.

10.7 The Generic Attribute Profile

The final piece of the attribute puzzle is the Generic Attribute Profile (GATT) procedures. The Attribute Protocol defines how a client and server can send standard messages between one another. The GATT procedures define standard ways that services, characteristics, and their descriptors can be discovered and then used. The GATT procedures can be considered to be split into three basic types:

- Discovery procedures
- Client-initiated procedures
- Server-initiated procedures

There is one additional type of procedure that doesn't fit into any of these groups. This is the Exchange MTU procedure that uses the Exchange MTU Request from the Attribute Protocol to determine the MTU size that is used for any subsequent messages. This procedure does not have to be used; therefore, the default MTU of 23 octets would be used.

10.7.1 The Discovery Procedures

There are four basic things that need to be discovered. First, the client needs to discover the primary services. Once the primary services have been discovered, all the other information on a server that is grouped with this primary service can be discovered. The client can then use the range of handles for each interesting primary service to discover the referenced secondary services, the characteristics that are actually exposed by this instance of the service. For each characteristic found, the set of descriptors can then be discovered. Only after all this is complete can the services be "used" by client and server-initiated procedures such as reading or writing characteristic values or descriptors.

10.7.2 The Discovering Services

There are three ways to discover services:

- Discover All Primary Services
- Discover Primary Service By Service UUID
- Find Included Services

10.7.2.1 Discover All Primary Services

When a client connects to a device and wants to find all the primary services exposed on the device to determine what it can do, it uses the Read By Group Type Request with the handle range set to 0x0001 : 0xFFFF and the attribute type set to Primary Service. The server responds with the one or more primary services that it finds. The response includes not only the handle of the service declaration but also the last handle for the attributes of this service. The response also includes the value of the service declaration so that the client can determine that it understands each service.

Unless the last handle of the last service is 0xFFFF or an Error Response was received, the client sends another Read By Group Type Request with the starting handle updated to be one greater than the last handle of the last service in the previous response. This way, the client can enumerate all the services on a device.

It should be noted that the Read By Group Type Response cannot return both services with 16-bit UUIDs and 128-bit UUIDs in a single response. Therefore, the server will return all 16-bit UUIDs before an attribute with a 128-bit UUID in one response, and then the 128-bit service in the next response, and the remainder of the 16-bit UUIDs in subsequent responses. This is optimized for the more common 16-bit UUIDs used by standard services. To help with this, 16-bit UUID–based services are recommended to have lower numbered handles.

10.7.2.2 Discover Primary Service By Service UUID

Sometimes, a client just wants to use a particular service and doesn't want or need to enumerate any other services. For example, a light switch would only need to discover the Light Service and wouldn't care about any other services that this device exposes. To help with these "simple clients," a special procedure is used that is optimized for discovering primary services that have a known type.

The client sends a Find By Type Value Request to the server with the handle range set to 0x0001 : 0xFFFF, the type set to Primary Service, and the value set to the service type that is wanted, for example, Light Service. The server responds with the handle ranges for each light service that is found.

Some services will be defined such that they are "singleton" services; they can only be instantiated once on any given server. For these services, the response will only ever include a single handle range. Other services will allow themselves to be instantiated multiple times; thus, the response will have multiple handle ranges, one for each service.

It might be possible—although unlikely—that the number of instances of a given service on a device will exceed the eight service handle ranges that can be included in a single response. In this case, the same updating of the starting handle that was used earlier is used to find the other services.

10.7.2.3 Find Included Services

Once the primary services are discovered, secondary services and other referenced services can then be discovered. This involves looking for an Include declaration by using the Read By Type Request. This time, the starting handle and ending handle would be set to the handle range of each service previously found. Typically, only two references can be returned in a single response. Therefore, the request needs to be sent with the starting handle set to one higher than the last handle returned.

Once referenced services have been discovered, the references to these services can also be discovered by using the same procedures.

10.7.3 Characteristic Discovery

Once the services have been discovered, the characteristics of each service can then be discovered. To discover characteristics, both characteristic discovery and characteristic descriptor discovery must be performed.

10.7.3.1 Discover All Characteristics of a Service

To perform characteristic discovery, Read By Type Request is used with the handle range set to the handle range for the service and the type set to Characteristic. This allows all the characteristic declarations to be discovered and read.

Within a service, the mandatory characteristics should be ordered first, followed by any optional characteristics. This allows a client that is looking for the one mandatory characteristic in a service to terminate this procedure early.

For each characteristic in a service, the characteristic declaration is returned, together with the handle of this characteristic. The declaration includes its properties, the handle of the attribute that contains the characteristic's value, and the type of this characteristic. This means that once you have discovered the characteristic declaration, you can determine what this characteristic represents, what you can do with this characteristic, and the handle for the subsequent read or write procedures.

10.7.3.2 Discover All Characteristic Descriptors

Once each characteristic's declaration has been discovered, it is then possible to find all the descriptors for each characteristic. This is done by using the Find Information Request with the handle range set to the handle range for each characteristic declaration grouped with the characteristic.

It is not possible to find the handles of all the attributes that are grouped with the characteristic declaration directly. It is possible to determine the handles of all the characteristic declarations within a service, and with this knowledge, determine the handles associated with a given characteristic. For example, if you have a service that has a declaration at handle 0x0100, and the end handle for this service is 0x010F, with characteristic declarations at 0x0102 and 0x0108, the attribute handles for the attributes grouped with the first characteristic of this service would be in the range 0x0103 to 0x0107, and for the second characteristic, they would be in the range 0x0109 to 0x010F.

The Find Information Response includes the handles and types of all the descriptors for the characteristic. Any characteristic descriptor that is not understood by the client can be safely ignored. Any characteristic descriptor that is understood by the client can be used to either understand the characteristic further, or configure

10.7 The Generic Attribute Profile

the behavior of the characteristic. For example, the client could use the Characteristic Presentation Format to understand how to display the value on the client's display. The client could use the Client Characteristic Configuration Descriptor to configure the characteristic to be notified or indicated.

10.7.4 Client-Initiated Procedures

There are four things that a client can do with a characteristic:

- Reading characteristic value
- Writing characteristic value
- Reading characteristic descriptors
- Writing characteristic descriptors

10.7.4.1 Read (Long) Characteristic Values

After the characteristics of a service have been discovered, the value of the characteristic can be read. The value is stored in an attribute that is pointed to by a handle in the characteristic declaration. The type of this attribute is also the same as the characteristic UUID from the characteristic declaration. This means that once the characteristic descriptor has been discovered by using the characteristic discovery procedures described earlier, the characteristic value can be read by using either a Read Request or a Read Blob Request.

The difference between using a Read Request or a Read Blob Request is very subtle and can cause confusion. Attributes can have a fixed length, and if this fixed length is less than the attribute protocol's MTU, the Read Request can be used to read the characteristic value. If the attribute has a fixed length but this length is longer than the attribute protocol's MTU, first a Read Request would be used to obtain the first 22 octets of the characteristic's value, followed by one or more Read Blob Request messages to obtain the remaining parts of the characteristic's value.

If the characteristic value can have a variable length, for example, a string, then it must be assumed that the value is longer, and then the Read Request followed by Read Blob Request messages must be used to read the complete value. This procedure can terminate early if the variable length value is shorter than expected. For example, if the characteristic value is 21 octets in length, only a single Read Request will be necessary; if the characteristic value is 22 octets in length, a Read Request followed by a Read Blob Request will be necessary, with the Read Blob Response containing no additional octets of value. Therefore, it is not possible to know if the Read Response or any Read Blob Response contains the last octet of the attribute value or is just full up because it has exactly the right length. Thus, the client keeps issuing a Read Blob Request until it receives a Read Blob Response containing less than 22 octets of data.

10.7.4.2 Read Using Characteristic UUID

There are some instances for which you just want to read the value of a characteristic, without having to first find all the characteristic declarations within a service and then read the value. For example, to read the battery level, it would be much more efficient to just request the value of the Battery Level characteristic. This can be done by using the Read By Type Request with the type set to the required characteristic UUID.

10.7.4.3 Read Multiple Characteristic Values

It is also possible to read multiple characteristic values at the same time. This does require that the handles of each of the characteristic's values are known. The biggest issue with this procedure is that each of the values must have a known size. This is because there are no frame boundaries on each value. The Read Multiple Request is used to perform this procedure.

However, even when taking this restriction into consideration, it can also provide some interesting possibilities. The last characteristic value requested can have a variable size because the size of this value is determined by the size of the attribute protocol packet.

10.7.4.4 Write (Long) Characteristic Value

To write characteristic values, the Write Request is used. This can only be used to write short characteristic values; by default, less than or equal to 20 octets in length. To use this, the characteristic value's attribute handle must have been discovered.

If the attribute value that is to be written is longer than 20 octets, a different procedure will need to be used. This involves using both the Prepare Write Request to prepare the long value to be written, followed by the Execute Write Request to actually write the value. It might appear to be overkill from a protocol point of view to use the same procedure to write long characteristic values as is used for reliable writes, but there is a very good reason for this. The Attribute Protocol is atomic in operation only for a single request. If the long attribute write occurred over multiple requests, causing the value to be partially changed between each request, any other device trying to read the value would possibly read an invalid value. By using the prepare write queue to prestore the long value to be written and then executing this write in a single Execute Write Request, the atomic nature of the write can be ensured.

10.7.4.5 Characteristic Value Reliable Writes

When attempting to write values with the maximum reliability, the Characteristic Value Reliable Writes procedure is used. This procedure can also be used for writing multiple characteristic values at the same time in an atomic operation. For example, when moving a machining tool to a new position in 2-D space, if the x and y values were written in sequential requests, two straight lines would be created, one horizontal and one vertical. However, if both the x and y values were prepared and then executed in a single atomic operation, a single line would be created.

The procedure uses the same Prepare Write Request and Execute Write Request that is used in the Write Long Characteristic Value procedure, with one additional check. For each value that is prepared, the Prepare Write Response is compared with the request to ensure that the handle and value in the response are the same as those used in the request. As explained in Section 10.6.13, this ensures the maximum security against single bit errors. If the handle and value are different, the Execute Write Request is used with the flags parameter set to cancel the prepare queue. Consequently, all the values that were prepared will have to be prepared again.

10.7.4.6 (Signed) Writing Without Response

Sometimes, a value needs to be written very quickly and a response is not required at the protocol level. This procedure uses the Write Command to send the value to be written to the characteristic value's attribute. There is no response with this command; therefore, it is assumed that if a response is required, it would be delivered through another characteristic being notified, or is out of band, or not required.

For example, when turning a light on, it is not necessary to receive a response from the peer device over the Attribute Protocol because the user will see the light turn on. Another reason for using this procedure is when the response might not be available immediately. For example, a time synchronization service might expose a characteristic that can be written without response to start the synchronization process. When the synchronization has completed, the newly updated time value can be notified to complete the process. The final reason for using this procedure is when the response would be too complex to fit into a single message. For example, a message server could expose the state of a single message and have a characteristic that can be written without response to change which message is exposed. There is no reason to have a response to the write because the value of the associated characteristics will have been updated by the time those values are read.

For some devices, the data that is written needs to authenticate the initiator of this message. To do this, the Signed Write Command is used. Again, this does not have a response, so it can be used in exactly the same situations as a *Write Command*, but with the additional security that authentication provides.

For example, the television would like to know that the remote control that just sent the "power off" message was the remote control that is bonded with the television. Typically, this would have required the full encryption of the link to ensure authentication. Encryption takes time to set up and requires many resources on devices to create cipher streams for encrypting each message. For some applications, the time taken to encrypt the link will push the latency requirements of the application outside the latencies that can be tolerated. The Signed Write Without Response procedure solves this. The message can be authenticated before the link is established, and then sent on an unencrypted link as the very first message on the link. This significantly reduces the latency for the message being transmitted to the other device but does not stop eavesdroppers from listening to this transaction. However, if this message is just to say "turn on" or "turn off" a device, there is no confidential information in this message.

10.7.4.7 Read/Write (Long) Characteristic Descriptors

Characteristic descriptors are not the same as the characteristic value, but the procedures to access them are very similar to the procedures used to read and write the characteristic values.

For reading the descriptors, the Read Request and Read Blob Request are used. For writing the descriptors, the Write Request and Prepare Write Request/Execute Write Request are used.

10.7.5 Server-Initiated Procedures

Not all procedures are initiated by the client. Some are initiated by the server, including notifications and indications. Typically, the client will configure the server to send these messages or cause the server to send the messages because it commanded the server to do something that generates these messages.

There are two types of GATT procedures that are server initiated:

- Notifications
- Indications

10.7.5.1 Notifications

A notification is a server-initiated message that can be sent at any time by a server to a client. These messages have no flow control mechanisms, so a client might not have enough buffer space to hold all the received messages and is allowed to drop them. It is useful to consider notifications as unreliable messages. Notification uses the Handle Value Notification message.

10.7.5.2 Indications

An indication is a server-initiated message that can be sent at any time by a server to a client. These messages have flow control and therefore a server cannot send an indication until the last indication was confirmed as received by the client. These indications are therefore considered to be reliable messages. Indications use the Handle Value Indication message as well as the Handle Value Confirmation message sent by the client to acknowledge the receipt of the indication.

10.7.6 Mapping ATT PDUs to GATT Procedures

- Exchange MTU Request is used in the Exchange MTU procedure in GATT.
- Find Information Request is used in the Discover All Characteristic Descriptors procedure in GATT.
- Find By Type Value Request is used in the Discover Primary Services By Service UUID procedure in GATT.
- Read By Type Request is used in the Find Included Services, Discover All Characteristics of a Service, Discover Characteristics by UUID, and Read Using Characteristic UUID procedures in GATT.
- Read Request is used in the Read Characteristic Value and Read Characteristic Descriptors procedures in GATT.
- Read Blob Request is used in the Read Long Characteristic Values and Read Long Characteristic Descriptors procedures in GATT.
- Read Multiple Request is used in the Read Multiple Characteristic Values procedure in GATT.
- Read By Group Type Request is used in the Discover All Primary Services procedure in GATT.
- Write Request is used in the Write Characteristic Value and Write Characteristic Descriptor procedures in GATT.
- Write Command is used in the Write Without Response procedure in GATT.
- Signed Write Command is used in the Signed Write Without Response procedure in GATT.
- Prepare Write Request and Execute Write Request is used in the Write Long Characteristic Value, Characteristic Value Reliable Writes, and Write Long Characteristic Descriptors procedures in GATT.
- Handle Value Notification is used in the Notification procedure in GATT.
- Handle Value Indication is used in the Indication procedure in GATT.

Chapter 11
Security

> *There are two types of encryption:*
> *one that will prevent your sister from reading your diary, and*
> *one that will prevent your government.*
> —*Bruce Schneier*

11.1 Security Concepts

Security is a complex subject that for many people is just a black box. If the technology is secure, that is enough for lots of people. However, there are many things about security that need to be understood in the context of Bluetooth low energy. These include the following topics:

- Authentication
- Authorization
- Integrity
- Confidentiality
- Privacy

11.1.1 Authentication

Authentication is defined as a way to prove that the device with which you are connecting is actually the device it claims to be and not a third-party attacker. This is done by using two basic methods:

- Initial authentication and sharing of a secret
- Re-authentication using a previously shared secret

For example, when a user opens a bank account, she must provide documentation to prove that she is who she says she is. This authenticates the user to the bank, typically because the bank trusts the issuers of these documents. Passports, identification cards, and other government-issued documents such as a driver's license are typically used for this. The bank then gives the customer a plastic bank card and a Personal Identification Number (PIN) that she can use at a later time to re-authenticate herself with the bank. When the customer wants to remove money out of this bank account she must authenticate herself to the machine by using the card and the shared secret PIN. Any person who has that card and that PIN can authenticate himself as the account holder to the bank—even if he isn't in actuality.

In Bluetooth low energy, authentication is performed in three different ways:

- During the initial pairing of devices, an authentication algorithm is used to authenticate the connecting device. This might involve entering a passkey into one or both devices. This allows the link to be encrypted, and any shared secrets that will be required later can be distributed. When these shared secrets are stored, the devices are said to be "bonded."

- When reconnecting to a device with which you have previously bonded, one of these devices can send a signed command to the other device to authenticate that it knows the shared secret that was previously distributed. The signature is created by using the shared secrets exchanged at bonding, and as such, it cannot be falsified by a third party. Part of this signed command must be a counter that is incremented for each message sent to prevent *replay attacks*.

- When reconnecting to a device with which you have previously bonded, either device can initiate encryption. Each and every data packet that is transmitted from then on will incorporate a message integrity check (MIC) value that authenticates the sender of that message to the other device by using the previously distributed shared secrets.

11.1.2 Authorization

Authorization is defined as the assignment of permission to do something. This is usually done in two ways:

- Documentation that provides authorization
- Authorization that is actioned directly

A concert normally has few, if any, authentication requirements, but authorization to enter the concert is typically enforced by using something called a ticket. These tickets are provided by the event organizer to the guest who presents the ticket at the concert as authorization to enter. There is no authentication that the person holding the ticket is the same person who authenticated the payment for these tickets; in other words, no photo identification is needed.

Another example of authorization is the instant approval to allow somebody to do something. For example, if you quickly lent somebody your computer and ask him only to use a single program, you are authorizing him to use that program and no others.

This same authorization model can also be used wirelessly. When something connects with your device, you can authorize it to access certain parts of your device, but not everything within it.

11.1.3 Integrity

Integrity is defined as the internal consistency and lack of corruption of data. When any data is sent from one device to another, either by using a wired or wireless communications protocol, the data is subject to introduced errors. These errors are important to detect and guard against.

From a security point of view, some errors could be introduced by a third-party attacker to attempt to change a valid message into a malicious message. For example, by replaying a captured message but changing one bit in a message that means "lock door" to "unlock door," the security of a building could be severely compromised. It should be noted that cyclic redundancy checks (CRC) are used to protect against bit changes, but these are typically too weak to be considered a security measure. It is too easy to change not only a few bits in the message but also a few bits in the CRC to match. To ensure integrity, a much stronger form of message authentication is required that also checks the integrity of the original message.

11.1.4 Confidentiality

Confidentiality is defined as the intent to keep something secret. The most common representation of confidentiality is in films when you see the characters handling company or government reports marked as "confidential." Unfortunately, this is not a good use of the word because anybody who can see that report can read it. In Bluetooth low energy, confidentiality means that even if a third-party eavesdropper receives a message, she cannot decode it. The process of enciphering a message is called encryption. The *enigma machine* developed during the second world war is a classic example of a device that could encrypt or decrypt a message.

11.1.5 Privacy

Another security concern that should be considered is how private any communication is. Complete anonymity is difficult to provide. Take, for example, a famous person boarding an aircraft; there are plenty of people who will recognize him just by his face. It is therefore almost impossible to be granted complete anonymity in every location.

Wireless communications should not make it easy to track somebody. If the devices that somebody carries are constantly allowing other parties to track his movement throughout a space, there could be some interesting and spooky consequences. For example, stores might give you special offers based on what you have bought in other stores. That's not so bad, right? But a stalker could seed an area with devices and automatically track you through that area, which is a less attractive prospect than getting a few discounts.

Thus, privacy is the ability to prevent others from recognizing you by the devices that you are carrying, and not to allow them to track your movement throughout a space.

11.1.6 Encryption Engine

Within Bluetooth low energy, there is a single cryptographic block that is used as a one-way function to generate keys and also to encrypt and provide integrity checks. This encryption engine is called the Advanced Encryption System (AES) as defined by the NIST publication FIPS-197.[1] Bluetooth low energy uses the 128-bit version of this, known as AES-128.

A generic way of looking at AES is to consider a single function, E, that takes both a key and some plain-text data, and results in a cipher-text data block. AES is therefore a block cipher. The key is 128 bits in length, the plain-text data is 128 bits in length, and the resultant cipher-text is 128 bits in length. This can be expressed as shown in Equation 11-1:

$$ciphertext = E_{key}(plaintext) \tag{11-1}$$

It is interesting to note that if the key doesn't change rapidly, the algorithm is very efficient. Each time a new key is used, lots of calculations need to be performed to set up the internal state of the encryption engine. After the setup is performed, each new plain-text value that is input can be quickly converted into cipher text. The security algorithms in Bluetooth low energy make use of this property of the AES encryption engine.

11.1.7 Shared Secrets

Virtually all security is based on shared secrets. You bank card's PIN is a secret that is shared between you and your bank. Your computer password is a secret that is

---

[1] You can find NIST FIPS-197 at http://csrc.nist.gov/publications/fips/fips197/fips-197.pdf

shared between you and either your computer or your company's information systems departmental computers. Your house key is a shared secret between the metal that you hold in your hand and the physical formation of the metal within the door lock.

Within Bluetooth low energy, there are many shared secrets known as keys. A key is just a shorthand way of saying "shared secret." There can be plenty of keys, just like you have a car key, a door key (or two or three), a bicycle lock key, or a key to access your work.

There are five main keys in Bluetooth low energy:

- Temporary Key
- Short-Term Key
- Long-Term Key
- Identity Resolving Key
- Connection Signature Resolving Key

11.1.7.1 The Temporary Key

The Temporary Key (TK) is used during the pairing procedure. It is set to a value that is determined by the pairing algorithm and used to calculate the Short-Term Key.

"Just Works" is a mode designed to make connection to Bluetooth low energy devices possible when very limited user interfaces prevent user entry or verification of pass key values. The TK value when using "Just Works" is zero. This means that there is no authentication being performed and therefore this connection and any keys distributed over it would be vulnerable to *man-in-the-middle* attacks.

"Passkey Entry" is a mode used when the user interfaces on both devices allow at least the display or entry of a number value. The TK value when using the "Passkey Entry" algorithm would be set to the numeric value that is to be input on both devices. This numeric value is a value from 0 to 999999. This means that there is a significant probability that a man-in-the-middle attack will not guess the same value that is being used by the connection. There is only a one-in-one-million chance that the right value is guessed. For authentication, this is a reasonable probability; therefore, a key generated using this algorithm is considered an authenticated key, protected from man-in-the-middle attacks.

The last TK value is when the "Out Of Band" algorithm is used. This is when both devices have information that has been acquired by using another technology than Bluetooth. For example, if Near-Field Communication (NFC) was used to transfer a value between the two devices, this value can be used as the TK value for authentication. A key generated by using out-of-band data is considered authenticated and protected from man-in-the-middle attacks because it is assumed that the out-of-band technology is also not subject to these types of attacks.

11.1.7.2 Short-Term Key

The Short-Term Key (STK) is used as the key for encrypting a connection the very first time two devices pair. The STK is generated by using three pieces of information: The Temporary Key is used as the key for the encryption engine, and two random numbers, S_{rand} and M_{rand}, are contributed by both the slave and master devices in the initial pairing request. S_{rand} and M_{rand} are concatenated with the | symbol, as illustrated in Equation 11-2 and then encrypted using the Temporary Key.

$$STK = E_{TK}(S_{rand} | M_{rand}) \qquad (11\text{-}2)$$

The contribution of random numbers by both the slave and the master increases the security of the whole system because any attacker can only contribute 64 bits of the 128-bit random value. It is much harder for a man-in-the-middle attacker to guess which one of 2^{64} possible values the peer device used.

11.1.7.3 Long-Term Key

The Long-Term Key (LTK) is distributed once the initial pairing procedure has encrypted the connection. The LTK can be a random number that is stored in a security database. It is also possible that the LTK is generated on the slave device.

Slaves are by design resource-constrained, so having a security database might be considered too much. To solve this problem, the slave also distributes two other values: EDIV and Rand. These two values are stored on the master and sent upon a reconnection to the slave. The slave can then calculate the LTK that should be used, or more accurately, the LTK that it had previously given to the master.

Upon reconnection to a previously paired and bonded device, the LTK is used to encrypt the link. This means that full pairing is not required each and every time a device connects.

11.1.7.4 Identity Resolving Key

The Identity Resolving Key (IRK) gives a device that knows a peer device's IRK the ability to resolve (work out) a peer device's identity. Privacy could be performed by always using a fully random address. These devices could change their random addresses at random times, and the device would not be trackable or even connectable by any trusted device. The problem is how to be both private due to using a random address, and also be identifiable by trusted devices.

To solve this problem, the IRK is used when generating the random address. This is done by splitting the address into two parts: a random part, and a hash of

11.1 Security Concepts

this random part with the IRK, as shown in Equation 11-3.

$$hash = E_{IRK}(rand) \qquad (11\text{-}3)$$

By placing both the random number and the hash into the address field, a peer device that knows the IRK can be checked to see if they match.

One way to think of this is to consider a private person who uses a different alias each time he talks to another person. To keep this simple, assume that instead of 2^{24} possible name pairs, we have just three. The first time he could be known as "Floella Benjamin," the next time as "Bob Hope," and the next time as "Charlie Dimmock." If you see the names Floella with Benjamin, or Bob with Hope, and so on, then you know it is that person. If the name were "Floella Hope" or "Bob Dimmock," it is not that person because "Floella" always comes with "Benjamin."

Therefore, it's possible for a device that has a list of IRKs for each separate bonded device to do an exhaustive match on each of these IRKs with the received private addresses. A match likely means that the correct device has been found.

It should be noted that there is not a one-to-one correspondence of devices to random numbers and hash values. There are approximately 70,000 billion possible fixed device addresses and, therefore, devices, yet there are only 4 million random numbers that can be used. Therefore, there is a reasonably high probability that for two given devices with the same random number and different IRKs, it would be possible to have the same hash value. This is why privacy is typically combined with authentication to ensure that it really is the correct device and not somebody who has the same combination of address parts.

11.1.7.5 Connection Signature Resolving Key

The Connection Signature Resolving Key (CSRK) gives a receiving device the ability resolve a signature and therefore authenticate the sender of the message. The CSRK is distributed from the source of the message to the destination device for the message. Once distributed over an encrypted link, the link can be disconnected. Upon reconnection, because the message only needs to be authenticated and not sent in a confidential matter, the message can be signed.

To sign the data, the CMAC function defined by NIST Special Publication 800-38B[2] is used. This uses the CSRK as the key for this function, along with a SignCounter. The SignCounter is a 32-bit value that must be incremented for each message from the source device to the destination device for the duration of the bond. The SignCounter is set to zero immediately after bonding and incremented for each new packet sent, regardless of whether the devices disconnected in the meantime.

---

2. You can find NIST Special Publication 800-38B at http://csrc.nist.gov/publications/PubsSPs.html

11.2 Pairing and Bonding

To enable most of the security features in Bluetooth low energy, two things must happen: The devices must pair with each other, and then once the connection is encrypted, they must distribute keys that can be used to encrypt, enable privacy, and authenticate messages. If these keys are saved for a future time, the devices are bonded.

Therefore, to understand how security works, it is essential to understand how the pairing and key distribution system works. It is also important to understand that the initial connection between two devices is different from the subsequent connections between the same two devices.

11.2.1 Pairing

Two devices that initially have no security but wish to do something that requires security must first pair with each other. Pairing involves authenticating the identity of the two devices to be paired, encrypting the link, and then distributing keys to allow the security to be restarted on a reconnection much more quickly the second time around.

Pairing has three distinct phases:

- Exchange of pairing information
- Authentication of the link
- Key distribution

11.2.2 Exchange of Pairing Information

The first phase of pairing involves the exchange of pairing information that is used to determine both how to pair the two devices and what keys are distributed during the last phase.

It is important to note that just because there can be very complex algorithms used during the pairing operation, how the devices pair is probably the biggest single opportunity to have the user reject the whole wireless technology. If pairing is difficult and complex for the user, there is a risk that the user will fail to pair the device the first time around and take the product back to the store.

Bluetooth low energy uses the same pairing process as that used for the Secure Simple Pairing feature in Bluetooth classic.

11.2 Pairing and Bonding

Each device first determines its input and output capabilities, together with other paring information. The input and output capabilities are selected from a list of possible capabilities:

- No Input No Output
- Display Only
- Display Yes/No
- Keyboard Only
- Keyboard Display

To determine which of these five possible values should be used, the device determines its input and output capabilities and feeds them into a matrix, which is shown following this paragraph. The input capabilities can be either "no input," the ability to select "yes/no," or the ability to input a number by using a "keyboard." The output capabilities can be either "no output" or "numeric output." Numeric output in this context means the ability to display a six-digit number.

	No Output	Display
No Input	No Input No Output	Display Only
Yes/No	No Input No Output	Display Yes/No
Keyboard	Keyboard Only	Keyboard Display

These five device input and output capabilities are communicated between the devices by using the *Pairing Request* message.

The *Pairing Request* message is sent as the first security message from a device. This contains not only the capabilities but also other pairing information, including a bit stating if out-of-band data is available, and what the authentication requirements are, if any. It also includes the list of the keys that are being requested to be distributed at the end of the pairing procedure. The authentication requirements include whether bonding is enabled and whether man-in-the-middle protection is required.

In response to this, a *Pairing Response* message or a *Pairing Failed* message can be sent.

The *Pairing Response* message includes basically the same information as the Pairing Request. However, if the request indicated that out-of-band data was present but the responder doesn't have any out-of-band data, then the *Pairing Failed* message is sent instead. In fact, the Pairing Failed message can be sent at any time during the pairing procedure to give devices the opportunity to fail the current pairing; for example, when a parameter doesn't match what is supported or expected.

Once the *Pairing Request* and *Pairing Response* have been exchanged, the two devices can then move to the second phase of the pairing procedure.

11.2.3 Authentication

Using the information from the *Pairing Request* and *Pairing Response*, the two devices can deterministically use the pairing algorithm. The two input and output capabilities are used in the following table to determine which algorithm is used:

	Display Only	Display Yes No	Keyboard Only	No Input No Output	Keyboard Display
Display Only	Just Works	Just Works	Passkey Entry	Just Works	Passkey Entry
Display Yes/No	Just Works	Just Works	Passkey Entry	Just Works	Passkey Entry
Keyboard Only	Passkey Entry	Passkey Entry	Passkey Entry	Just Works	Passkey Entry
No Input No Output	Just Works	Just Works	Just Works	Just Works	Just Works
Keyboard Display	Passkey Entry	Passkey Entry	Passkey Entry	Just Works	Passkey Entry

It should be noted that for some combinations, it is entirely possible to lead the users into doing the right thing. For example, if device A has a Display Yes/No capability while device B has a Keyboard Display capability, then device A can display a six-digit number, and the user of device B can see that number and type it into device A. That value can then be used for the TK value.

Once the TK value has been determined, a really simple but hard-to-attack procedure is used to help stop a man-in-the-middle attack. A random number is generated by each device, and a confirmation value is also calculated based on that random number, the TK value, the known values of the pairing so far including the device addresses, and the parameters from the pairing request and response messages see Equation 11-4.

$$Confirm = E_{TK}(Random|PairingMessages|Addresses) \qquad (11\text{-}4)$$

This value confirms that all the known parameters and addresses used so far are the same on both peer devices. This protects against man-in-the-middle attackers.

Both devices exchange the random numbers and the confirm values; therefore, they can check that the confirmation values match the random numbers and all the other shared information. The interesting twist is that the confirmation values are sent before the random numbers are exchanged. By doing this, an attacker would have to guess which one of the 2^{128} possible random numbers the peer device would use to calculate the confirmation value, before it knows what that random number might be.

If the confirmation values do not match the random numbers, then a *Pairing Failed* message would be sent to terminate the pairing because something was wrong. If the confirmation values do match the random numbers, both devices have the same input parameters from the pairing request and response, the same address information, the same TK value, and the correct random numbers.

Assuming that everything is confirmed correctly, the random numbers exchanged during the authentication are then used to calculate the STK value as described in Section 11.2. This STK value is then used to encrypt the link by using the Link Layer encryption procedures, as described in Chapter 7, The Link Layer, Section 7.10.3.

11.2.4 Key Distribution

Once the connection is encrypted by using the STK, it is then possible to distribute the required keys. These keys are distributed one at a time because at 128 bits in length, they only just fit into a single packet.

The following keys can be distributed:

- LTK
- IRK
- CSRK

The LTK is distributed along with EDIV and RAND because the slave does not have a security database. The slave can use these bits of information to generate the LTK for that master directly.

Both the master and slave can distribute all the types of keys. This is because for the current connection, the topology might be that device A is a master and device B is a slave; however, in subsequent connections the topology might be reserved such that device A is the slave and device B is the master. It is therefore possible to distribute the LTK from the master to the slave so that if they reconnect in a different topology, they can still reconnect quickly.

A slight issue that must be considered is that the addresses used during the pairing might not be the actual address of the device; the addresses used during pairing can be either a random or a private address. It is therefore also useful to distribute the identity information of the two devices so that this information can be used as the unique value in the database. This allows future connections to be performed by using this identity information rather than a random address that could be out of date.

11.2.5 Bonding

Bonding is really a Generic Access Profile discussion; however, it is useful to consider its operation at this point also. Bonding is nothing more than the saving of the keys and associated identity information in a security database. If the device does not save these values, the devices will have paired but not bonded. If one device saves them but the other doesn't, upon reconnection, only one device will have the LTK, and thus the starting of encryption will fail.

To avoid this prospect, both devices exchange bonding information during the initial pairing so that they know whether the other device saves this bonding information. If the other device does not save the information, then once the attempt at starting encryption fails, the hosts will attempt to pair again.

11.3 Signing of Data

When a device is connected but not encrypted, it is possible to send data that is authenticated without confidentiality. To do this, a CSRK is exchanged during the very first connection at which pairing occurred. After this point, as long as no data is exchanged that requires confidentiality, signing can be used.

To sign the data, the CMAC algorithm is used. This algorithm takes the message to be authenticated, a SignCounter, and the CSRK used to authenticate the sender and then generates a signature value.

The message that is authenticated is typically an attribute protocol Signed Write Command and consists of the opcode, handle, and value that is being written.

The SignCounter is a 32-bit value that must be incremented on each new data packet that is sent. This is used to stop replay attacks. If the same SignCounter is transmitted in a subsequent data packet, it must be assumed that an attacker has received the previous message and is attempting to replay the message in the hope that it might do something interesting. Unfortunately for the attacker, the receiver just discards any message with a SignCounter that is less than the next expected value. It should be noted that the peer device needs to store this next expected SignCounter value for each device with which it is bonded if signed data is supported.

The SignCounter value must be included in the message that is sent because the receiver doesn't know if the next message received has the next expected SignCounter. Consider the case when this is used to open and close the garage door. The message that commands the garage door to open or close is not confidential; anybody watching from the street will see the car turn up, followed by the garage

11.3 Signing of Data

door opening. If you go out one day far away from the garage and press the garage door remote control button a number of times, each time it is pressed a new message needs to be generated. By the time you return home and press the button to actually open the garage door, the SignCounter value might be higher than what is expected. The garage door itself won't consider this a threat because the value is not one that has been previously used.

Chapter 12

The Generic Access Profile

> *It is no coincidence that in no known language does the phrase "As pretty as an Airport" appear.*
> —*Douglas Adams*

The final part of the core specification is the Generic Access Profile (GAP). This defines how devices can discover and connect with one another and how they bond. It also describes how devices can be broadcasters and observers and, as such, transfer data without being in a connection. Finally, it defines how the different types of addresses can be used to allow private and resolvable addresses.

12.1 Background

One of the most important things to understand with Bluetooth low energy is how two devices first find one another, work out what they can do with one another, and how they can find and connect with one another repeatedly. This is really what GAP defines. To illustrate how Bluetooth low energy works, let's consider a typical user scenario:

A user has just returned home from the store from which she has just purchased a new heart-rate belt. She already has a low energy–enabled phone and wants to connect the heart-rate belt to the phone. She has been buying a lot of low energy devices recently; the list includes a low energy–enabled television with remote control, a low energy lighting system, and, of course, low energy sensors in her computer, shoes, and watch. How does she configure the heart-rate belt to work with her phone?

The user opens the heart-rate belt box and reads the instruction leaflet that tells her to remove the plastic tab from the heart-rate belt to turn the device on. It then instructs her to discover the heart-rate belt. On her phone, she accesses the Bluetooth menu, taps the "Add Devices" button, and then watches as a number of devices appear on the screen.

At the top of the list of devices is the heart-rate belt she just purchased along with an icon of a heart-rate belt. She selects the heart-rate belt, and the phone moves to its application store and displays a list of applications that work with this device. Some applications are free, some cost money, and some are from the same brand as the heart-rate belt. She selects an application and installs it on the phone. A few seconds later, the application is running and displaying heart-rate information on the screen. The user goes out for a run to test the application.

The next day, the user puts on the heart-rate belt again and starts the application she downloaded yesterday. Again, the heart-rate information is displayed on the screen. It's almost like magic—she simply uses the application, and the device works.

A few days later, one of the user's friends recommends a different application that he downloaded for his own heart-rate belt. She goes to the application store, searches for this application, downloads it, and then runs it. The application again displays the heart-rate information, but now also includes additional information about how hard the user is working out. The application is using the same heart-rate information from the belt, but this time it's using the data in a different way.

This example shows how a user can simply begin by using a Bluetooth low energy device, and how the flexibility of Bluetooth low energy services means that they are not tied to a single application to use that device.

12.1.1 Initial Discovery

To discover a device, you must scan for devices that advertise. Advertising devices transmit packets to any scanning devices by using a many-to-many topology. The problem with this is that every device that is connectable, but not necessarily discoverable, will be scanned. To solve this, in addition to data, some flags are also broadcast that reveal whether the device is discoverable or connectable.

There are two types of discoverability. The first type is used by devices that are "limited-discoverable." This is used by devices that have just been made discoverable on a temporary basis. For example, the first time a device is powered on, it would be limited-discoverable, or if the device had a button that allowed it to be discoverable temporarily. A device is not allowed to remain in the limited-discoverable mode for very long. This is because limited-discoverable mode is intended to allow devices to stand out from the crowd of general-discoverable devices. If devices stayed limited-discoverable for a long time, they would not stand out. Therefore, a device can only be limited-discoverable for about 30 seconds.

The second type is used by devices that are "general-discoverable." This is used by devices that are discoverable but have not been recently engaged in interaction. For example, a computer that is discoverable, such that other devices can find and connect to it, but has not recently had this discoverability turned on, would be generally discoverable. A device that is searching for other devices would typically place generally discoverable devices lower down the list of found devices because

12.1 Background

these are probably not going to be as immediately important to the user as those limited-discoverable devices are now.

Determining device discoverability is the combination of scanning for all devices and filtering on the discoverable flags that each of these devices is broadcasting. It is also possible to use additional filters when presenting the list of devices to a user. For example, the proximity of the device can be used to present the devices that are closer to the searching devices at the top of the list and those that are farther away at the bottom of the list. To do this, the transmit power at which the advertising packets are transmitted is compared with the received signal strength to calculate the path loss of the communication. Devices with a smaller path loss will likely be closer than devices with a larger path loss.

Another possible form of filtering involves using the service information in advertising data to segregate based on what a devices does. For example, if the user is looking for a heart-rate belt, then the fact that one device in the area supports this particular service probably indicates that this is the device with which the user wants to connect. There are two types of service information that a device can expose: a device can expose a list of the services that it has implemented, and a device can expose a list of services that it would like a peer device to support.

Therefore, with nondiscoverable devices, limited-discoverable devices, general-discoverable devices, path loss range filtering, and service-based filtering, an intuitive interface can be made for the user. If the user interacts with a device, it appears at the top of the list; if he moves the device closer, it appears nearer the top of the list; if the device supports the services that are being searched for, they will appear closer to the top of the list. A user just performs the search, looks at the item at the top of the list, and, with high probability, connects to this correct device.

One little wrinkle in the device discovery procedure is name discovery. Users don't like to look at hexadecimal 48-bit numbers; instead, they prefer to look at user readable and understandable names. To allow every device to have a name, the GAP defines a characteristic that exposes the device name. The device name can also be included in the advertising data or the scan response data for devices.

To obtain the scan response data, active scanning needs to be used. Therefore, when discovering devices to display names on the screen, active scanning is typically required. This is because the device name is static data that would normally be included in the scan response data and not the advertising data. However, some devices have so much information they need to broadcast that they are unable to include the complete device name in the advertising or scan response data; instead, they include just part of the name or none of the name.

To obtain the complete device name in this situation, a connection must be made with this device and the device name read. Thankfully, the device name has the well-known characteristic type of "Device Name," which you can read by using the simple attribute protocol Read By Type Request.

12.1.2 Establishing the Initial Connection

Once the list of devices has been found and a device has been selected, the next step is the initial creation of a connection to the device. This initial connection is performed by initiating a connection to the same device address that was found from the advertising packets. Once the devices are connected, the connecting device performs either an exhaustive enumeration for all the services and characteristics of that device, or it looks only for the service or services that it is interested in, and the characteristics of those services.

For example, a phone or computer that has an application store would enumerate all the services that the device exposes. This service information can then be sent to the application store, and any applications in that store that support those services would be presented to the user. So, instead of the user having to use the application that came with a device, an ecosystem of applications becomes available. Some would be free, some would cost money, some would be made by the manufacturer of the device, and some would be from independent software companies.

The alternative method is that the device only performs service discovery for one or possibly a very limited list of services. For example, a television that is connected to a remote control would only search for the human interface device service and possibly the battery service, but nothing else. This means that even if the device supported more services than these two, the television would never even perform service discovery on them. This method would typically be used by devices that have a limited user interface or by devices that are connecting to peripherals that only use a very simple set of functionality.

The result of service discovery is a list of those services that a device exposes. The client can then use these services. In the application store model, it is the application that takes the next step of characteristic discovery and configuration. Characteristic discovery is just like service discovery in that a device can either enumerate all the characteristics within a service or just use the well-known characteristics that it knows a service must expose. For example, the battery service must expose the battery level characteristic. So, if it doesn't need to discover any characteristics within the battery service, a client can read the battery level characteristic directly.

12.1.3 Service Characterization

For a heart-rate belt, the characteristic discovery and configuration might be a little more elaborate. For example, the heart-rate service might expose just the heart rate, or an aggregation of the heart rate and the time intervals between heartbeats, or just the time interval between heartbeats. It can also expose other information that it has calculated from the heart-rate sensor. A client can pick and choose which

12.1 Background

characteristics to read. Some clients might only be interested in the heart-rate value, whereas others might want to read all the information.

For the efficient transfer of data between devices, notifications or indications should be used. To configure the characteristics for which a client wants to receive notifications or indications, the client must write the client characteristic configuration descriptor. This descriptor enables the required functionality. The device then starts sending these notifications or indications whenever necessary. For the heart-rate example, a client might configure the heart-rate value to be notified, and for that service, the value would be notified once each second, even if the heart rate doesn't change. For other services, the time interval can be more flexible. The battery service, for example, might only notify a value when the value changes.

Thus, the application can configure the set of characteristics to be notified or indicated and then wait until the service sends this data through. This means that even if the user changes the heart-rate application, the new application will continue to receive the heart-rate notifications that were previously configured.

12.1.4 Long-Term Relationships

Most of the time, a peripheral is associated with a single central device. Your proximity tag is associated with your phone; your keyboard with your computer; your garage door opener with your garage door. When one device is associated with another, they are are said to be *bonded*. Bonding is a two-step process by which two devices that barely know one another authenticate themselves and share secrets.

For bonding to be successful, both devices also need to be configured to be bondable. A device that is not bondable—perhaps because it is already bonded with another device and can only manage a single bond at a time—doesn't need to advertise that it is bondable.

When both devices are bondable and one of these devices wants to bond, the first step of the bonding process has started. After this, the input and output capabilities of the two devices are exchanged, an authentication algorithm is chosen based on these capabilities, and the devices authenticate one another. This results in an STK that is used to encrypt the link (for more information about this, go to Chapter 11, Security, Section 11.1.7.2).

With the link now encrypted, the second step of the bonding process can be performed. This step involves exchanging shared secrets that can be used when reconnecting to the same device. These shared secrets are typically keys that have various uses: a Long-Term Key (LTK) for encrypting subsequent connections, an Identity Resolving Key (IRK) for resolving private addresses, a Connection Signature Resolving Key (CSRK) for checking the signature of signed attribute protocol commands, as well as the distribution of some identification information sent by the slave to the master so that the slave doesn't have to store the information.

12.1.5 Reconnections

Sometimes, devices will discontinue a connection. This might be because they've already sent everything they needed to send and don't want to waste energy maintaining a connection. A light switch will create a connection, send the "turn on the light" command, and then quickly disconnect. Sometimes, the connections might be maintained for significantly longer periods of time. For example, a keyboard might remain connected with a computer until the computer is turned off, at which point, it disconnects. When the computer is turned back on again, it needs to reconnect to the keyboard.

Reconnections are both easy and hard. In Bluetooth low energy, all devices that want to be slaves in a reconnected connection need to be advertising by using connectable advertising events. It might not be discoverable, or it might be either limited or general-discoverable, or it might not allow connections from any device. Also, for a device to connect to an advertising device, the master must scan or initiate a connection to the device with the particular address that is advertising. This means that the reconnecting device must be in the white list of the scanning or the initiating device if white lists are being used (for more information about white lists, go to Chapter 8, The Host/Controller Interface, Section 8.4.10).

12.1.6 Private Addresses

Some complications are introduced if the device that is advertising is using private addresses. A private address is a random address that changes periodically, for instance, once every 15 minutes. This means that even if you discovered a device that is advertising now, you will not be able to determine if that same device is around in 20 minutes' time, because it might be using a different address. This can at first appear to be an impossible problem to solve.

The solution to this problem is a three-step process. The first step is to save an IRK during bonding; the second is to use this key to generate a resolvable private address; and finally, the master must scan for all devices and resolve these private addresses by using all the IRKs and only connect to devices that it believes it has identified.

A resolvable private address is a type of random address that comprises three parts. The first part is a set pattern of two bits to identify that this random address is a resolvable private address. This reduces the computational load on the scanning device so that it only attempts to resolve private addresses on resolvable private addresses. The second part is a 22-bit random number. The third part is a hash of the random number with the IRK that was shared during bonding.

The combination of the random number and the IRK means that each private-addressed device effectively has four million possible addresses that fingerprint it.

A master that wants to reconnect to a slave device that is using private addresses must therefore take each resolvable private address that it scans, check with each and every IRK for each device that it knows could be using private addresses, and then compute the hash value that would have been used for each of these devices, given the random number. If this hash value matches the value used in the resolvable address, then there is a fairly high probability that this is the device identified by that key.

This is not absolutely certain. It's possible that another device with a different IRK with the same random number generates exactly the same hash value. However, a quick connection and encryption can quickly check that this is the correct device. The IRK and the LTK used for encryption or the CSRK used for signature authentication will be different for each of these devices, which can quickly confirm that this is either the right device or the low probability that a duplicated hash value has happened and it is the wrong device.

Private addresses have some disadvantages. The biggest one is that the host of the master must perform brute-force checking of each and every IRK for each resolvable private address that it receives. If that host knows many private devices, this could take some time. The HCI Encrypt command is very useful in this instance, especially if the host has few computation resources available.

Another disadvantage is that white lists cannot be used to make connections easily. The only way to connect to private devices is to first scan for the resolvable private addresses, compute if this is a private address of one of the devices to be connected to, and then connect to it manually. This increases the power consumed by the host because the host must perform many computations to resolve the addresses.

12.2 GAP Roles

There are four GAP roles defined for a Bluetooth low energy device:

- Broadcaster
- Observer
- Peripheral
- Central

A broadcaster is a device that sends advertising packets. Typically, this is used to broadcast some data from a service to other devices that happen to be in an observer role. A broadcast must have a transmitter but does not need a receiver. A broadcast-only device, therefore, only needs a transmitter.

An observer is a device that scans for broadcasters and reports this information to an application. An observer must have a receiver; it can also optionally have a transmitter.

A peripheral is a device that advertises by using connectable advertising packets. As such, it becomes a slave once connected. A peripheral needs to have both a transmitter and a receiver.

A central is a device that initiates connections to peripherals. As such, it becomes a master when connected. Just like a peripheral, a central needs to have both a transmitter and a receiver.

A device can support multiple GAP roles at the same time. For example, a device can be a broadcaster and a peripheral at the same time.

12.3 Modes and Procedures

Within GAP, there are two basic concepts that are used to describe how a device behaves. These are modes and procedures.

When a device is configured to behave in a certain manner for a long time, this is known as a mode. If a device is configured to perform a single action that has a finite time period over which this behavior will occur, this is called a procedure.

For example, when a device is broadcasting, this is called "broadcast mode." Broadcasting typically lasts a long time; perhaps that is the single purpose of the device. When a device is looking for broadcasters, this is called the "observation procedure." Observations typically occur for a very short period of time to build a user interface or find specific information that is needed.

Within GAP, the following modes are defined:

- Broadcast mode
- Nondiscoverable mode
- Limited-discoverable mode
- General-discoverable mode
- Nonconnectable mode
- Directed-connectable mode
- Undirected-connectable mode
- Nonbondable mode
- Bondable mode

12.3 Modes and Procedures

Within GAP, the following procedures are defined:

- Observation procedure
- Limited-discovery procedure
- General-discovery procedure
- Name-discovery procedure
- Auto-connection establishment procedure
- General-connection establishment procedure
- Selective-connection establishment procedure
- Direct-connection establishment procedure
- Connection parameter update procedure
- Terminate connection procedure
- Bonding procedure

To understand GAP is therefore to understand how these modes and procedures interact with one another. For example, the broadcasting mode and observation procedure can be logically combined to allow observations of broadcasters.

12.3.1 Broadcast Mode and Observation Procedure

When a device is in broadcast mode, it is using the Link Layer advertising channels and packets to transmit advertising data. This broadcast data can be observed by devices by using the observation procedure.

When broadcasting, the advertising data must be formatted correctly, as described in Section 12.5.

It should be noted that some devices might have only a transmitter and are therefore broadcast-only devices. For these devices, it would not generally be possible to use private addresses or signed data in the broadcast data. Both of these require knowledge of keys that are distributed during the bonding procedure that require a connection, which require both transmitter and a receiver. Of course, if these keys could be distributed out of band, then they could still be used; however, there is no standard defined method for this to occur.

12.3.2 Discoverability

In GAP, only devices operating in the peripheral role are discoverable. Devices that are trying to discover these devices will be in the central role.

A peripheral device can be in one of three different modes used for discoverability: nondiscoverable mode, limited-discoverable mode, and general-discoverable mode. Discoverability, in the context of GAP, does not mean that a device is advertising, but that the device is discoverable by a user interface on a peer device. A device can transmit advertising packets without being discoverable, as understood by GAP.

Therefore, it's necessary to include discoverability information as part of the advertising data for devices to be discoverable in the context of GAP. This differentiates them from the devices that are advertising or broadcasting data that are not discoverable. To do this, the Flags AD information (see Section 12.5.1) includes two bits that are used to determine if a device is in nondiscoverable mode, limited-discoverable mode, or general-discoverable mode.

12.3.2.1 Nondiscoverable Mode

A peripheral that is in nondiscoverable mode cannot set either the limited-discoverable mode or general-discoverable mode bits in the Flags AD information. If no other bits in the flags information are set, then the Flags AD does not have to be included in the advertising data. Nondiscoverable mode is the default discoverable mode; therefore, the host must command a change from this default to one of the other discoverable modes.

12.3.2.2 Limited-Discoverable Mode

A peripheral that is in limited-discoverable mode sets the limited-discoverable mode bit and clears the general-discoverable mode bit in the Flags AD. Limited-discoverable mode is used by a device with which the user has recently interacted, for example, the user just turned the device on for the first time or pressed the connect button, or the user just made the device discoverable in some user interface.

Limited-discoverable mode is only allowed to be used for approximately 30 seconds. This means that if another device finds a limited-discoverable device, it can be fairly certain that this is a peripheral with which the user has very recently interacted. Therefore, this device is most likely to be the one to which the user is trying to connect at this moment.

Because limited-discoverable mode is primarily designed for devices that are trying to be discovered and enter a connection, it is highly recommended to also include the following useful information in the advertising data that will help build the user interface:

- Tx Power Level AD to allow sorting of devices found by path loss and therefore range
- Local Name AD to allow the name of the device to be displayed
- Service AD to allow filtering based on what the device supports

Limited-discoverable devices should advertise at a reasonable advertising interval to both allow the user interfaces to be populated quickly and also to reduce the time required by the master device for scanning for Bluetooth low energy devices. The recommended interval should be somewhere between 250 milliseconds and 500 milliseconds. Obviously, a device that is discoverable and wants to connect should allow connections from any device, and, as a result, white lists would not be used.

12.3.2.3 General-Discoverable Mode

A peripheral that is in general-discoverable mode sets the general-discoverable-mode bit and clears the limited-discoverable mode bit in the Flags AD. A device uses general-discoverable mode when it wants to be discoverable.

When compared with limited-discoverable mode, general-discoverable devices are very similar, except in the following ways:

- General-discoverable devices can be discoverable for an unlimited period of time; limited-discoverable mode has a 30-second maximum time limit.
- General-discoverable devices have a slower recommended advertising interval of between 1.28 seconds and 2.56 seconds; limited-discoverable devices are between 250 milliseconds and 500 milliseconds.

It should noted that general-discoverable devices will not be found very quickly. This is by design; it is always useful to populate a user interface with those devices that are in limited-discoverable mode first and then place those devices that are generally discoverable lower down the list.

The same transmit power-level information should still be included in the advertising data, along with the other user interface building information, to allow the priority sorting and filtering of the discoverable devices to be used by the connecting device.

12.3.2.4 Discoverable Procedures

A central device that wants to find discoverable peripherals either uses the limited-discovery procedure or the general-discovery procedure. These two procedures are essentially identical except for filtering they perform based on the Flags AD information.

If there is no Flags AD information available in an advertising packet, or this information exists and neither the limited-discoverable mode bit nor the general-discoverable mode bit is set, then the advertiser is nondiscoverable and would not be discovered.

If the Flags AD information has the limited-discoverable mode bit set, the device is always discovered. This is true regardless of whether the limited-discovery procedure or the general-discovery procedure is used.

Table 12–1 Discoverability

	Limited-discoverable mode bit set	General-discoverable mode bit set
Limited-discovery procedure	Discovered	Ignored
General-discovery procedure	Discovered	Discovered

If the Flags AD information has the general-discoverable mode bit set, the peripheral is only discoverable when using the general-discovery procedure. These peripherals are not discovered by the limited-discovery procedure.

As shown in Table 12–1, the general-discovery procedure finds all discoverable peripherals regardless of whether they are limited or generally discoverable, whereas the limited-discovery procedure finds limited-discoverable peripherals.

12.3.3 Connectability

In GAP, only devices operating in the peripheral role use connectable modes. Devices that are trying to connect to these devices are in the central role and use connection establishment procedures.

As with discoverability, connectable peripherals can be in one of three modes: nonconnectable mode, directed-connectable mode, and undirected-connectable mode. However, connectability is more complex for the central device because there are four different connection establishment procedures: auto, general, selective, and direct.

From the GAP perspective, connectability is much easier to manage than discoverability. The Link Layer provides two types of connectable advertising packets: ADV_IND and ADV_DIRECT_IND; and two types of nonconnectable advertising packet types: ADV_NONCONN_IND and ADV_SCAN_IND. Thus, it's always possible for the host to use the correct type of advertising packet based on the connectable mode it is using.

12.3.3.1 Nonconnectable Mode

A device that is in nonconnectable mode cannot use a connectable advertising packet type when it advertises. This means that it can only use the ADV_NONCONN_IND or ADV_SCAN_IND advertising packet types. This mode is the default mode, so the host must perform an action to make a peripheral device connectable.

12.3.3.2 Directed-Connectable Mode

A peripheral device that wants to connect very quickly to a central device uses the directed-connectable mode. This mode requires the use of the ADV_DIRECT_IND

12.3 Modes and Procedures

advertising packets and, as such, cannot be combined with the discoverable modes because these advertising packets have no host-generated advertising data.

Because directed advertising packets are sent very quickly, this mode can only be used for a maximum of 1.28 seconds, after which the controller will automatically stop advertising. The host should not immediately start directed-connectable mode after it has timed out because this can severely restrict the ability for other devices to broadcast, be discoverable, or establish connections. Therefore, it's recommended that if directed-connectable mode times out, undirected-connectable mode should be the fall-back mode.

When using the ADV_DIRECT_IND advertising packets, both the current peripheral's device address and the central's device address are included in the packet. This means that the peripheral must have at least been connected with this device previously to know the device address. So, it's assumed that this mode would only be used when devices are bonded.

12.3.3.3 Undirected-Connectable Mode

A peripheral device that is connectable but does not need to establish a connection very quickly or wants to be connectable while saving as much energy as possible would use the undirected-connectable mode. This mode requires the use of the ADV_IND advertising packets. Because ADV_IND advertising packets can include the Flags AD information, a device can be discoverable at the same time as being in undirected-connectable mode.

A peripheral stays in undirected-connectable mode until the host either moves to nonconnectable mode or a connection is established. Therefore, as soon as the new connection is terminated and the device wants to continue to be connectable, it will move back into the undirected-connectable mode again. Obviously, a device that was connectable but is now in a connection cannot still be connectable because this would require that the Link Layer support two connections in the slave role at the same time, and this is not a supported state for the Link Layer.

12.3.3.4 Auto-Connection Establishment Procedure

The auto-connection establishment procedure is used to initiate a connection to many devices at the same time. A typical user scenario for Bluetooth low energy is that a central's host is bonded with a number of peripherals and it wants to establish a connection with each of these devices as soon as they start to advertise. For example, a central device might want the ability to establish a connection to any sensor device that has a new reading when the device starts advertising.

To allow a central device to make a connection to many devices at the same time, the host must first populate the white list with the set of devices that should be connected and then initiate a connection to the white list (for more information about white lists, go to Chapter 8, The Host/Controller Interface, Section 8.6.1). Typically, this would be the set of bonded devices about which the host is aware. Once one of these devices is found, because it is using directed-connectable mode or undirected-connectable mode, a connection is established. If other peripherals still need to be connected, the auto-connection establishment procedure would be restarted.

There are two downsides to this procedure: the procedure only has one common set of connection parameters that can be used, and this procedure cannot connect with devices that are using private addresses.

Because the connection parameters are determined by the central's host when it initiates a connection to the white list, it is not possible to have different connection parameters for different peripherals. This, therefore, works best when the types of peripherals are reasonably similar, at least from a connection parameter point of view. Obviously, it is possible to change the connection parameters once the connection has been established. So, it might be necessary to initially use a fairly compromised set of connection parameters when using this procedure.

Private addresses also cause severe problems for this procedure. Because private addresses must be changed frequently and randomly to provide privacy, it is almost impossible to predict the private address that a device will be using at any given point in time. As a result, it's not possible to use this procedure to connect to peripherals that are using private addresses.

12.3.3.5 General-Connection Establishment Procedure

In an attempt to solve the problems outlined in the preceding section, the general-connection establishment procedure does things slightly differently. Instead of using white lists, this procedure uses passive scanning to find all the devices that are advertising.

For devices that are using resolvable private addresses, these addresses are checked against all known IRKs for the devices to which the central devices want to connect. If the address resolves, the host stops scanning and uses the direct-connection establishment procedure by using the resolvable private address that it received.

Obviously, if a device address was received that was not a resolvable private address but is instead in the list of peripherals to which it wants to connect, it stops scanning and uses the same direct-connect establishment procedure by using the known device address.

The downside of this procedure is that there is a time gap between the host discovering that a device is advertising and the time when a direct connection establishment starts. This means that at a minimum, the connectable peripheral must send two connectable advertising packets before a connection can be established. This is a natural consequence of using private addresses.

This general-connection procedure also requires the processing of all advertising packets received by the controller in the host, even if the advertising packets are not connectable advertising packets or are not from devices that are interesting to the central device. This can require significantly more energy consumption in the host than the auto-connection establishment procedure.

12.3.3.6 Selective-Connection Establishment Procedure

The selective-connection establishment procedure is used to initiate a connection to many devices at the same time, but for which each device has different connection parameters. This solves the single-connection parameter problem from the auto-connection establishment procedure.

To perform the selective-connection establishment procedure, the host places the set of devices that are to be connected in the white list and then starts scanning by using this white list. This means that only peripheral devices that are in the white list and are advertising are passed up the host; all other devices that are advertising in the area are filtered out immediately by the controller.

When the host receives the advertising information from the controller for a device in the white list, it can first check that this device was using a connectable advertising packet type. If it was, the host stops scanning and uses the direct-connection establishment procedure to initiate a connection to this specific device. Because each peripheral might want to have a different set of connection parameters, the host can also look up the desired connection parameters for the peripheral that can be used by the direct-connection establishment procedure.

This has the same downside as the general-connection establishment procedure in that it takes a minimum of two advertising packets to establish a connection, but it solves the problem of having the same connection parameters as the auto-connection establishment procedure. Unfortunately, this does not solve the privacy issue of the auto-connection establishment procedure regarding the resolution of private addresses.

12.3.3.7 Direct-Connection Establishment Procedure

Many of the previous procedures reference the direct-connection establishment procedure. Basically, this procedure is used to establish a connection to a single, specific device by using a set of connection parameters.

It does not use the white list; instead, it initiates a connection directly to a single device address (for more information, go to Chapter 8, Section 8.6.2). This is why the general and selective-connection establishment procedures reuse this procedure to make the actual connection, once they have found the device address of the device that is advertising.

12.3.4 Bonding

Just like discoverability and connectability, bonding defines modes and procedures, except that there are only two modes: nonbondable mode and bondable mode; and just one procedure: the bonding procedure.

12.3.4.1 Nonbondable Mode

The default mode for devices is the nonbondable mode. This means that a device will not accept bonding at this time. No keys will be exchanged or stored by the device.

12.3.4.2 Bondable Mode

If a device wants to be bondable, then it is in bondable mode. When a device is in bondable mode, it will accept a bonding request from a peer device. To be in bondable mode, the bonding bit is set in the authentication requirements of the *Pairing Request* message during pairing (for more information on pairing, go to Chapter 11, Security, Section 11.2.2). If possible, the device exchanges security keys and stores them if necessary.

12.3.4.3 Bondable Procedure

If a device wants to bond with a device it believes is bondable, it uses the bondable procedure. When using the bondable procedure, the device initiates pairing with the same bonding bit set in the authentication requirements of the *Pairing Request* message.

Therefore, for bonding to work, the device that uses the bondable procedure initiates pairing with the bonding bit set. If the peer device is bondable, it will respond with the bonding bit set. If all this happens, the keys will be distributed after the link is encrypted, and the keys are then stored. Once the keys have been distributed and stored, the devices are bonded.

12.4 Security Modes

When pairing, the algorithm chosen can determine if the pairing performed a strong authentication to establish that the peer device was what it said it was, or whether it was not possible to provide a strong authentication. For example, a device that has no input or output capabilities will not be able to prove that it is actually the device that was connected because it has no way to authenticate itself. This is known as an unauthenticated pairing.

12.4 Security Modes

Other devices such as keyboards and computers can prove their identity by transmitting some information out of band that can be used for authentication. The computer can show a six-digit number on the computer screen and then ask the user to type this number into the keyboard. Both devices now have knowledge of this six-digit number, but any device that was eavesdropping on the connection cannot determine this shared number by just listening to the packets being transmitted. This is known as an authenticated pairing.

The strength of the authentication algorithm used is also used to label the keys that are distributed during bonding. If a device was paired by using an unauthenticated algorithm, the keys distributed during bonding must also be labeled as unauthenticated keys. If a device was paired by using an authenticated pairing, the keys will be labeled as authenticated.

Any subsequent reconnections using these labeled security keys would again be labeled with the strength of the algorithm implemented during the initial pairing. It is therefore possible to describe an encrypted connection as being encrypted by using either an unauthenticated pairing or an authenticated pairing, even if this pairing occurred on a previous connection.

12.4.1 Security Modes

GAP defines two security modes with up to three levels of security. The first security mode is for different levels of encryption within a connection; the second security mode is for different levels of data-signing protection. These security modes are used when describing what level of security is required for a given service. For example, some services have have no security requirements, whereas others might want to have as much security as possible to protect against eavesdroppers and ensure that authentication and confidentiality requirements are met.

The following security modes and levels are defined:

- Security Mode 1 Level 1: No security
- Security Mode 1 Level 2: Unauthenticated pairing with encryption
- Security Mode 1 Level 3: Authenticated pairing with encryption
- Security Mode 2 Level 1: Unauthenticated pairing with data signing
- Security Mode 2 Level 2: Authenticated pairing with data signing

Security Mode 1 Level 1 is used when there are no security requirements between two devices. This is the default security level on a link.

Security Mode 1 Level 2 is used when data confidentiality is required but authentication is not required or was not possible. It is possible to send data that requires no security over a link that is encrypted to this level of security.

Security Mode 1 Level 3 is used when data confidentiality and authentication are required. This is the strongest security mode and level. It is possible to send data that requires a lower security level over a connection that is currently using this level of security.

Security Mode 2 Level 1 is used when neither data confidentiality nor authentication is required.

Security Mode 2 Level 2 is used when data confidentiality is not required but authentication is required for this data. It is possible to send data that does not require authentication over a link that uses authentication.

Table 12–2 demonstrates that sometimes it is possible to send the data for a different security mode and level on a connection that is stronger than required. For example, a connection that is encrypted by using an authenticated pairing can send

Table 12–2 Security Modes and Levels

	Unencrypted Link and Unauthenticated Pairing	Unencrypted Link and Authenticated Pairing	Encrypted by Using Unauthenticated Pairing	Encrypted by Using Authenticated Pairing
Security Mode 1 Level 1	Send data	Send data	Send data	Send data
Security Mode 1 Level 2	Encrypt then send data	Encrypt then send data	Send data	Send data
Security Mode 1 Level 3	Not possible—must repair by using an authenticated pairing	Encrypt then send data	Not possible—must repair by using an authenticated pairing	Send data
Security Mode 2 Level 1	Sign data using unauthenticated key	Sign data by using authenticated key	Send data	Send data
Security Mode 2 Level 2	Not possible—must repair by using an authenticated pairing	Sign data by using authenticated key	Not possible—must repair by using an authenticated pairing	Send data

any type of data. For example, an unencrypted link that has authenticated keys from a previous pairing can be used to send data that requires Security Mode 2; however, it would require the link to be encrypted before data can be sent that requires Security Mode 1 Level 2 or Security Mode 1 Level 3.

12.5 Advertising Data

Whenever a device transmits advertising packets, the advertising or scan response data has a defined format. The format is just a sequence of advertising data structures. Each structure starts with a length field, which determines how many more bytes of data are part of this structure. Immediately after the length is the advertising data type. This is normally just 1 byte in length, but could be 2, 3, or more bytes if necessary. A device that doesn't understand a given advertising data type can just ignore it and skip to the next structure. Any additional data bytes within the structure are defined by the data type.

For example, the TX Power Level data type defines that the data, after the length and data type, is a single byte in length. This means that to output the TX Power Level in an advertising or scan response packet requires 3 bytes in total: one for the length, one for the data type, and one for actual power level.

It is possible for some advertising data to be variable length. The local name of a device could be anything from a few bytes to many bytes in length. The length field at the start of the data structure, however, binds the number of bytes that are being output in the advertising data. Therefore, it is not necessary to include any terminating bytes. It is also possible to not include the complete value of a given value, especially if that would overflow the packet size or remove something else just as valuable from the packet.

Most data that could be expected to overflow is typically represented by multiple data type values: one for the complete value if possible, and another for the shortened version. This way, the receiver can determine that the value is not complete, and thus, it can attempt to find the rest of this data through some other means. For example, the local name might be longer than is possible to include in a short advertising packet, but it is also possible to read the complete name by using the GAP's Device Name characteristic.

12.5.1 Flags

The Flags AD is a sequence of bit fields that can be any length, from zero to many bytes long. Any bytes that are not included in the advertising data are assumed to have the value zero. This means that the flags field can be extended with additional flags bits as necessary. A device that receives a Flags AD that is longer than it expects can just truncate the data without problem.

The following flags are defined for Bluetooth low energy:

- Limited-discoverable mode
- General-discoverable mode
- BR/EDR Not Supported
- Simultaneous LE And BR/EDR To Same Device Capable (Controller)
- Simultaneous LE And BR/EDR To Same Device Capable (Host)

The limited and general-discoverable mode flag bits are the same bits that are described in Section 12.3.2.

The BR/EDR Not Supported flag bit is used to notify a peer device before making a connection that it cannot make a connection by using Bluetooth classic; instead, it must use Bluetooth low energy. This is important because a dual-mode device cannot make a connection to another dual-mode device by using Bluetooth low energy. Therefore, a dual-mode device must check this bit to determine how it should initiate a connection to the device.

The Simultaneous LE And BR/EDR To Same Device Capable flag bits—one for the controller and one for the host—are used to determine if the peer device can initiate a connection over Bluetooth low energy if a Bluetooth classic connection already exists to that device. The controller and host might have different capabilities, and, therefore, both bits must be examined before two connections to the same device are attempted.

12.5.2 Service

There are multiple, varied types of service advertising data types. This advertising data type exposes a list of service UUIDs, one for each service. UUIDs come in two different sizes: 16-bit values or full 128-bit values. Some devices will not want to expose all the services that they support, or the complete list of services might be too long. So, both a full list of services and a partial list of services must be supported. Therefore, it is necessary to expose four different service advertising data types:

- Complete list of 16-bit service UUIDs
- Partial list of 16-bit service UUIDs
- Complete list of 128-bit service UUIDs
- Partial list of 128-bit service UUIDs

12.5.3 Local Name

The local name advertising data type comes in two variants:

- Complete Local Name
- Shortened Local Name

If the local name is too long to fit into the advertising packet, the Shortened Local Name data type will be used. The local name is a UTF-8 string, possibly truncated.

12.5.4 TX Power Level

The transmit power level advertising data type is used to expose the transmit power level used to transmit this information in the advertising packet. It consists of a single byte of data delineated in dBm.

12.5.5 Slave Connection Interval Range

The slave connection interval range exposes the preferred connection interval range for any subsequent connection to this peripheral. The master is in complete control of the connection parameters, and these parameters are typically set before the directed connection establishment procedure is used. Thus, this information is useful to allow the central to at least determine a range of connection intervals that it should use when connecting to the peripheral.

The interval range consists of two 16-bit values. The first value is the minimum connection interval; the second value is the maximum connection interval. Both values are in the same units as those used in the HCI command used to initiate a connection.

12.5.6 Service Solicitation

Sometimes a peripheral wishes to use a service on a central device. Unfortunately, the central device might not want to advertise, or the peripheral might not be designed to spend energy scanning for all possible central devices. The solution to this is to include a list of services that the peripheral would like the central device to support. This solicitation of services gives a central device that is looking for peripherals the ability to connect, and it can use this information to determine which peripherals are more likely to include the client functionality for one of its services.

The service solicitation advertising data consists of a partial list of either 16-bit service UUIDs or 128-bit service UUIDs.

12.5.7 Service Data

When a service is configured to broadcast, it uses the service data advertising data type. The first two bytes of the data are the 16-bit UUID for the service that is broadcasting the data. Any additional bytes are the actual service data that is being broadcast.

12.5.8 Manufacturer-Specific Data

The last advertising data type is manufacturer-specific data. The first two bytes of the data are a 16-bit company identifier for the ensuing data, followed by data that is specified by that company. This means that any company can define its own advertising data structures and expose them in an advertising packet.

12.6 GAP Service

GAP defines its own generic attribute profile service, the Generic Attribute Profile, which provides a device with a way to ascertain information about the device, including its name, what it looks like, and how to connect to it. The service exposes up to five characteristics:

- Device Name
- Appearance
- Peripheral Privacy Flag
- Reconnection Address
- Peripheral Preferred Connection Parameters

12.6.1 The Device Name Characteristic

The Device Name characteristic is a UTF-8 string that exposes the name of the device. Interestingly, a device can only have one Device Name characteristic, so it's possible to use the Read By UUID Request to quickly read the device's name without having to perform service discovery or characteristic discovery first.

12.6.2 The Appearance Characteristic

The Appearance characteristic is a 16-bit value that enumerates what the device looks like. Bluetooth Classic had a class of device field that could be discovered during the inquiry procedure, but this was a mix of what the device did and what it looked like. This has proven to be a real problem because some manufacturers filtered

12.6 GAP Service

on bits in the class of device, creating interoperability problems. For Bluetooth low energy, the chosen solution was to just concentrate on what was the most important information to build up a user interface rather than attempting to do filtering as early as possible. Therefore, the appearance is only allowed to be used to drive the icon that is displayed on the user interface next to the device.

The value in the characteristic is an enumerated value that is defined in the assigned numbers document, which is a living document. New values can be added to this enumeration when required by just requesting an assignment to be made.

12.6.3 The Peripheral Privacy Flag

The Peripheral Privacy Flag is used to expose if the device is currently using privacy. It is both readable and writable; this is where the use of this characteristic becomes confusing.

- If this characteristic does not exist, this device does not support privacy and will always use the public Bluetooth address.

- If this characteristic does exist but it is set to "disabled", again, this device is not using a private address and therefore will use the public Bluetooth address.

- If this characteristic exists and it is set to "enabled", it will always advertise by using a private address. However, that doesn't mean that it will be connectable by using a private address.

- If the Reconnection Address characteristic exists, the address used when making a connection will be this reconnection address, and not the advertising address.

This means that only when the reconnection address exists and the privacy flag is set to "enabled" is this device truly able to use private reconnections.

A central device might not want to use privacy with a device, or it might want to use privacy with a device that currently has the Peripheral Privacy Flag set to "disabled". In either case, the central device can attempt to write this characteristic. However, this write might not succeed.

If the peripheral has more than one central device bonded with it, having the second or later central device being able to enable privacy would mean that the first bonded central device would not be able to reconnect. Reconnecting would fail because it would be attempting to reconnect to the public address of the device and not the value it last read in the Reconnection Address.

Disabling privacy can also cause similar problems. This direction is not as dysfunctional as the central device that still thinks that privacy is enabled on the peripheral because the central at least can detect that the peripheral is using its public address in advertising packets, allowing it to reconnect to this address. Upon reconnecting, it can check the value of the Peripheral Privacy Flag to confirm that privacy has been disabled.

Therefore, the only safe way to use privacy is to have it always enabled from the start of the peripheral device being connected. It is not useful to constantly change the privacy state.

12.6.4 Reconnection Address

As explained in the preceding section, the Reconnection Address is used when privacy is enabled. This is the address that a peripheral device uses when it attempts to reconnect to a central device that it is bonded with, and the central device knows that the peripheral has privacy "enabled".

When used, the Reconnection Address will be a nonresolvable private address. Therefore, the only devices that can make a mapping from this reconnection address to a peripheral are the devices that are bonded with the peripheral and have read the reconnection address.

The reason that the reconnection address is part of the system is to allow for a power optimization when privacy is used. If there were no reconnection address, all private peripherals would advertise by using a resolvable private address. The central would then need to check every received advertising packet with a resolvable private address against every IRK that it knows. This is an expensive operation, typically because the host must get involved. By using a static nonresolvable private address, the value is static and can therefore be placed into the white list to allow the controller to efficiently establish a connection to the peer device without using the power-hungry host.

The interesting thing about the reconnection address is that for it to be considered private, it must be changed on each reconnection. Therefore, each time that a central device reconnects to a peripheral using the reconnection address, it will write a new reconnection address into this characteristic before doing anything else.

There is a possibility that the connection fails during this write request. In this case, the peripheral might have received the write request and not been able to send back the write response. Alternatively, the peripheral might not have received the write request. Therefore, the central must assume that the peripheral could be using the old reconnection address or the newly written reconnection address. It will therefore have to place both of these addresses in its white list.

12.6.5 Peripheral Preferred Connection Parameters

Many peripherals have been designed for a single use case and have a preferred set of parameters with which they work optimally. Instead of having the central device

12.6 GAP Service

guess what these parameters are, the peripheral exposes them in the Peripheral Preferred Connection Parameters characteristic. The central can then read them on the initial connection and quickly change the connection parameters to something that the peripheral really likes. Also, when reconnecting to a peripheral for which the central has remembered the parameters, it can place the appropriate values into the connection request, removing any need to change the connection parameters after the connection has been established.

Part IV

Application

Chapter 13, Central, describes how an application can use Bluetooth low energy to interact with peripherals.

Chapter 14, Peripherals, explains how a peripheral can be designed to make the best use of the Bluetooth low energy technology.

Chapter 15, Testing and Qualification, explains how to qualify a design so that it can be sold or distributed.

Chapter 13

Central

One is and is not in the centre of the maelstrom of it all.
—*Harold Pinter*

The application of Bluetooth low energy can be broken into two separate disciplines: the design of applications that find and interact with peripherals, and the design of peripherals that can provide information to these applications running on a central device. Both disciplines require knowledge of how all the previous parts fit together to make a functioning whole. This chapter looks at the applications from a central point of view. The next chapter subsequently deals with how a peripheral can be designed.

13.1 Background

Central devices are vitally important for Bluetooth low energy to be useful. Typically, these devices are highly functional and have complex user interfaces. This chapter will not go into how to write an application for a particular type of device; that is the job of that platform's developer program. Instead, important considerations will be given on how to optimize the user experience and the best ways to save power.

13.2 Discovering Devices

The first thing that any central device needs to do is to discover other devices. To do this, it can use either passive or active scanning. For passive scanning, a central device passively listens to any advertisement packets that peripherals are transmitting. Active scanning is when the central device, after hearing a peripheral, asks for more information.

If the central device is only looking for what devices are around, and perhaps any information they may also contain, it should use passive scanning. If the central device is also populating a user interface such as a screen or a window, active scanning should be used because the additional information can be useful to build the list of discoverable devices on a user interface.

The information that can be found by scanning includes the name of the device and a unique number that identifies the device. This identity can be used if the central device later needs to connect to it. It is also possible to find some broadcast data within the scan responses so that information that is being broadcast by a service can be obtained. This could be useful information such as the battery level or the current time.

Some information is not immediately available but can be obtained by making a quick connection to the device and reading it as needed. For example, the complete name of the device might be very long and might not fit into a single advertising or scan response packet. Therefore, it would be necessary to make a connection with the device to read the rest of the device name. When doing this, the application on the central device should be very careful about only reading characteristics within the correct service.

Therefore, first the application should perform a service discovery of the required service, for example, to look for the GAP Service. It can then look within that service for the Device Name characteristic value. This can be read directly using the Read Characteristic Value by UUID procedure. The GAP Service specification requires that only a single GAP service can exist on a device, and it will have only one device name characteristic value. If the device name is longer than what can be read in a single attribute protocol response, additional requests for the rest of the device name value can be made by using the handle that is returned in the first response.

This same procedure can be used for other useful information, such as the Appearance characteristic value in GAP.

It should be noted that when passively or actively scanning, not only can the application obtain the contents of advertising packets but it can also receive the received signal strength (RSSI) of these packets in the controller. This RSSI value can be subtracted from the Tx Power that might be included within the advertising packets to give a very basic estimate of the *path loss* and therefore an estimate of the distance between this device and the central device, as illustrated in the following equation:

$$path\ loss = TxPower - RSSI$$

If the *path loss* is very small—between 0 and 20—this indicates that the device is very close. If the *path loss* is very large, for example, above 70, the device is very far away. Because of the nature of wireless transmissions, these values should be averaged over a number of seconds so that any multipath interference is averaged out. Once this is done, the list of devices presented to the user should place the closest (the smallest *path loss*) at the top of the list. This way, the user will preferentially see those devices that are very close. When there are lots of devices, the user will need to scroll down to find those devices that are farther away.

Another consideration is that some devices might be advertising but not discoverable. Unless the user interface is displaying the broadcast service data, the devices that are not discoverable should be removed from the list of found devices.

13.3 Connecting to Devices

In the preceding section, the list of discoverable devices was created. This list can be presented to the user so that he can select a particular device (or devices) with which he wants to interact. The next task will be connecting to the device.

Connecting to a device selected from a user interface should just be a case of initiating a connection to that device. If, however, the chosen device is using a private address, care should be taken to recover gracefully if the device found has recently changed its private address. If that does happen, the initiation of the connection should timeout and the list of discoverable devices should be refreshed. The device might still be there, but it's using a different private address because it wants to protect its privacy. This is not a problem once the devices have bonded and exchanged Identify Resolving Keys (IRKs) because the central device can automatically refresh the list, resolve the private addresses to identify which one belongs to the desired device, and then connect to it.

When initiating a connection, a set of connection parameters need to be chosen. The parameters used depend on what the two devices are intending to do. Typically, peripherals have a Client Preferred Connection Parameters characteristic that gives a very strong hint to a central device about the types of connection parameters it prefers. When making the very first connection with a device, this information will not be available, and therefore a compromise between low power consumption and rapid characterization of a device needs to be struck.

The best set of parameters are those that have a fairly rapid connection interval, allowing for a rapid exchange of Attribute Protocol and Link Layer control messages during the initial connection. A reasonably large slave latency should also be offered so that the peripheral device can save power whenever possible. For example, a connection interval within the range of 15 milliseconds and 30 milliseconds, with a slave latency of 150 milliseconds, allows for both rapid collection of data about the peripheral using up to 60Hz connection interval and a 6Hz idle frequency for the slave.

The slave might request different parameters from those that the application on the central device chose. The central device should always try to honor these requests, especially after it has finished reading all the data that it wants at that moment.

It is always possible for the central device to change the connection latency again, once it has more data to send or receive.

13.4 What Does This Device Do?

After connecting to a peripheral device, the central device will want to know what the device does. To gather this information, it uses four procedures, in a specific order: primary services discovery, relationship discovery, characteristic discovery, and descriptor discovery.

The first procedure is the primary services discovery. These are the services that describe what the device does. For example, if the device has a battery, primary services would expose the Battery Service; if the device has a temperature sensor, it would expose the Temperature Service; if the device had a temperature sensor within the battery, this secondary temperature would not be exposed through a primary service because this might confuse the central device. Primary services only expose the primary functionality of a device.

Next, for each primary service that the central device knows could include another service, these relationships need to be discovered. These relationships could be because of an extending, combining, or reusing relationship.

The set of services that a device has does not necessarily determine the set of profiles that a peripheral device supports. There is no way to quickly determine the set of profile roles that a device supports. Instead, a more complex algorithm has to be used, matching the profile roles that the central device supports with the set of services that are exposed on the peripheral and checking which roles are valid. This could be a nontrivial operation; however, the devices that would be doing these checks have plenty of resources, and this type of complexity is not considered a Bluetooth low energy issue.

The benefit of this approach is that future client profile roles that use a set of services on a peripheral don't need to be designed into the peripheral when it is manufactured. This becomes a very flexible and extendable system ideally suited to the downloadable application models being deployed in many central devices.

Once the services have been discovered, the set of characteristics and their descriptors can also be discovered. There are no version numbers in Bluetooth low energy services; therefore, the only way to know if a given optional feature exists is to check for the exposure of a given characteristic that is linked to this optional feature. Alternatively, characteristic properties and descriptors such as the notification and Client Characteristic Configuration descriptor can be used to differentiate optional behavior.

13.5 Generic Clients

It is possible to build entirely generic clients. These are clients that can read and display characteristic values, possibly in a human-readable format. This gives central devices that have no understanding of the meaning of the individual services or characteristics on a peripheral device the ability to make them available to the user. There are two levels of generic clients: those that use the available information on the peripheral device directly, and those that augment this information with characteristic information available via the Internet.

The first level of generic clients finds all the characteristics within a peripheral and filters out any that cannot be read directly. It then also filters out any that do not have a Characteristic Presentation Format descriptor. This descriptor includes most of the information needed to change the binary representation of the data into a human-readable value. It does this by using the format and exponent fields to determine how to convert the value from a fixed-point value into a more intuitive number. The unit field is a UUID that encodes the unit of this value.

There are units for most physical quantities, which are taken from the BIPM[1] list of units. All standard SI units and most common units are included. Finally, a namespace:description pair is included in the descriptor to allow for an even finer-grained client display. This could allow not only for the weight of the item being measured to be converted from a 16-bit unsigned value into pounds weight, but also noting that this weight is from a hanging weighing machine as opposed to a vehicle weight bridge or a set of bathroom scales.

The second level of generic clients can display the most complex values in a characteristic. The Characteristic Presentation Format descriptor is limited in that it can only represent a very small subset of all possible data structures. Therefore, the second level of generic clients does not rely on that descriptor. Instead, it uses the knowledge that each and every characteristic type has a unique number, the UUID, that can be looked up on the developer website of the Bluetooth Special Interest Group (SIG)[2] to find an XML representation of the data format.

For every characteristic type in any service specification, there must also be an XML file defined for that characteristic. These files are used to help validate devices when they are being tested, but they can also be used to determine the structure of the characteristic value. The XML files can encode every possible data representation

---

1. Bureau International des Poids et Mesures. You can find more information at http://www.bipm.org/
2. Use https://developer.bluetooth.org/ to discover all possible characteristics.

required, including enumerated types, bitfields, optional fields based off the value of an enumerated value or a bitfield, and binary and decimal exponential fixed-point formats. These fields can also be concatenated together to make very complex data structures.

A generic client with a connection to the Internet can therefore find a readable characteristic value and perform a simple query of a website to download the characteristic representation's XML file. This file can be used to display the value to the user.

13.6 Interacting with Services

Once the central device has determined that it is necessary to interact with a service on a peripheral device, it makes a connection to that device and starts to read and write characteristic values and descriptors. The protocol used to do this, Attribute Protocol, is essentially a stateless protocol.

The protocol has no state when connected or when between one connection and the next. There is no "session protocol" either. To get around this limitation, all state is maintained at the application layers, where the applications can make intelligent decisions on how to save energy. The next sections describe how this is done.

13.6.1 Readable Characteristics

The most basic of services simply exposes a set of readable characteristics. For example, the Device Information Service contains one or more characteristics that provide additional information about a device. These basic services are easy to use. For each characteristic in the service that the client understands, it reads the value, either in a single request if the value is short, or by using multiple requests if the value is long.

13.6.1.1 Writable Characteristics

The next level of complexity is a service that has a characteristic that is both writable and readable. The Link Loss Service is a good example of this type of service. It has a single writable characteristic, Alert Level, that the client can write to configure the behavior when the link between the two devices is lost.

If the client writes "No Alert" into this characteristic, when the devices disconnect, the server does nothing. If the client writes "Mild Alert" into this characteristic, when the devices disconnect for any reason, the server will use a mild alert to the user to notify her of this occurrence. If the client writes "High Alert" into this characteristic, when the devices disconnect, the server will use as many bells, whistles, flashing lights, and other "alert" methods that it has at its disposal.

Some people who have experience using connection-oriented protocols can be worried by this. How do you "start" this service? How do you "stop" this service?

The key element to understand here is that the state of the service is exposed in the service characteristics. The Alert Level characteristic in the Link Loss Service determines the device behavior. If this holds the value "No Alert", the server will do nothing. You could consider this the "not-connected" state. If this holds the value "Mild Alert" or "High Alert", the server will do something when the client disconnects. You could consider that when this service's characteristic holds one of these values, it is "connected".

Of course, this implies that if the client wants to gracefully disconnect from the server that has a mild or high alert level, it must change the alert state by writing "No Alert" to this characteristic before it disconnects.

Many people will also ask about what this means if a server can have more than one connection at the same time, and one client writes the value "Mild Alert" to the link loss service's alert level characteristic, but the other doesn't and then disconnects. This is not a problem. The value for each characteristic of each service can be different for each client. If client A writes "Mild Alert", this does not mean that the server will alert when client B disconnects because the value for client B is still "No Alert".

An important difference between this type of service and those described in the following sections is that these services have characteristics that can be readable and writable. This means that a client can always check the current state of these services without having to remember what it had done previously. This is most useful when the client application unexpectedly terminates, perhaps during debugging of the client application software. When restarting, the client can just refresh its knowledge of the state by reading the appropriate characteristic values.

13.6.2 Control Points

Another type of service is one that holds no state, but the client can still write values to the service. This might appear strange at first, especially considering the services described in the preceding section were holding state. How can a service have a characteristic that is writable and not hold on to that state? The answer is easy: The service uses the value written immediately, and the server does not have any need to store that value after it has been consumed. This type of characteristic is called a *control point*.

The previously described Link Loss Service had an Alert Level characteristic that could be written that determined the behavior when the two devices disconnected. But what if the client wants to just make the server alert now? It could write the appropriate value into this service and then disconnect. But this is incredibly disruptive to any other applications that are also using other services on this device that might need the connection. Is there a better way? Of course there is, and it uses a control point characteristic.

The service is called the Immediate Alert Service and the characteristic is called Alert Level. Yes, it is the same alert level characteristic that was used in the Link Loss Service. But characteristics are simply a data format, in this case, an enumeration of three values, "No Alert", "Mild Alert", and "High Alert." The behavior is determined by the service, not the characteristic.

In the Immediate Alert Service, the Alert Level characteristic is only writable, and it causes an alert immediately. Because the alert is immediate, this characteristic cannot have a state. Any value written is immediately consumed, used to make an alert, and not stored. Therefore, there is no point in making this characteristic readable. The characteristic has no state.

There is another type of control point that is discussed in the Notifications and Indications subsection that follows.

The advantage of this type of control point-based service is that it doesn't matter which slave commanded the control point. Instead, the device will act upon the command written into the control point.

13.6.3 State Machines

The next type of service exposes a few writable control points along with one or more readable characteristics. These expose the state of a state machine. A state machine in this context is a "machine" that has an exposed state and a way of internally or externally changing states. Essentially, the only difference between a state machine and a control point described earlier is that the state is remembered in a state machine. This state can therefore be read and notified to the client if it changes.

To help illustrate this, let's consider a state machine for time synchronization. It would have the current states: the machine is doing nothing, and the machine is busy trying to find a more accurate value for the current time. Let's label these states "Idle" and "Searching." This is the exposed state of the state machine.

Next, we need a way for the device to control the state machine. This is done by using a control point. This is a control point as described just a moment ago, except that the control point is connected with the state machine that has an external state. For example, with the time synchronization service, this could be enumerated with two commands: Start Synchronization and Cancel Synchronization. This has a number of advantages.

Any device can interrogate the current state of the state machine through the machine's exposed state characteristic. This means that if three devices all want to synchronize time at the same time, they can all check the state and only command it to start a new synchronization if the machine is in the idle state.

More realistically, each of the devices can just command the state machine to start synchronization. This should not be a problem with a well-defined state machine that can continue to operate even when given potentially invalid commands. For example, if the state machine stays in the searching state when it was commanded to start synchronization, it would be safe to have multiple clients all ask it to start synchronizing any time they want. This is much preferred over the alternative of sending error responses to all clients when any sends a second start synchronization command, and then having to instigate some form of random back-off procedure to start synchronizing again.

The state machines must be sufficiently robust that all commands will have a defined state transition. For example, if the time synchronization service was idle and a client asked it to cancel synchronization, the state machine could take this as an error, or it could accept that the client was unaware that it was already in the idle state and do nothing. This latter approach is typically required in state machines, so all commands should be safe for the client at any time.

However, this scenario points up an interesting side effect: When one device asks for synchronization to start, it has no guarantee that synchronization will complete because it might be cancelled by another device. As an alternative, a service could be defined in which synchronization would always run to completion once started, unless canceled by that client.

13.6.4 Notifications and Indications

Services expose state information. Some of this data can change rapidly or randomly. It would be inefficient to constantly poll the state of a service on a device just in case the value has changed. Take, for example, a battery in a device. Sometimes, when the device is not being used, the battery level might only change once a day. However, if the device is actively being used, the battery level can change once every 15 minutes or more. How often should the client check the battery level? If the client checked the battery level once each day, then at the end of one day the client might think the battery is full, but the battery is actually at 4 percent. If it checked the battery level every 15 minutes, even when the device is idle, the battery level might drop faster than it should due to excessive energy use from reading the battery level.

Instead, it is possible to set up a characteristic to tell the client when the value changes. This way, the client is only notified when the state has actually changed. The client can wait for the changes in state to arrive from its peripheral devices and then efficiently deal with these changes.

To do this, the client can configure a characteristic to send these notifications as required. Most notifications are sent as defined by a service, but some can be configured further by using additional characteristic descriptors.

It should be noted that some characteristics don't support notification; only characteristics that have the correct properties and have a Client Configuration Descriptor will support notification.

It should also be noted that there is a second way of receiving data from a service. Notifications are sent at any time and aren't acknowledged by the application layer. Therefore, they must be considered unreliable. This is fine for many things, such as a status update on a state machine characteristic. But if it is a blood glucose reading, the concept of the server unreliably sending that data is untenable. Therefore, a client can configure characteristics to send indications. These are the same as notifications except that a confirmation message is sent back to the server to inform it that the client has received the data and that the application has received the data.

Indications are configured in the same way by using the Client Configuration Descriptor.

13.7 Bonding

Some devices need a longer relationship than only a single connection. Other devices might want to transmit data confidentially or to only transmit data to a device that has been authenticated to really be the same device as used last time rather than any available device in the area. This is achieved by using a security model that fundamentally results in a "bond."

A client that wants to establish a long-term relationship with another device will first connect to that device, find some services that it can use, and then initiate a secure connection with the device. These secure connections first authenticate that the device is the correct device. Next, it encrypts the link to ensure confidentiality. Finally, the devices exchange some pairing information. This is the critical bit: If this pairing information is stored on the client, the client has a "bond" with that device.

This is important because when the client reconnects to this device, it does not need to reauthenticate and exchange pairing information again. It just encrypts the link by using information it has stored as part of this bond, and the devices then have an authenticated and confidential data connection.

Bonding also provides other benefits. When devices are bonded, the server will save their configuration data for this client. This model allows a device that reconnects to immediately receive notifications without the need to reconfigure the server.

For example, if the central device configures a characteristic such as a battery level to be notified, the battery level would be notified to the client if it changes when it is connected. But if the client disconnects from the server, the server will not be able to send the notifications. With bonding, when the client reconnects to the server, the notification can be sent immediately.

This has a side effect that must be considered. The server could ignore all state changes for bonded devices and require the client to query the current state after it reconnects. If the data changes infrequently, this can be inefficient for both the client and server. This model also assumes that the client doesn't disconnect and reconnect frequently because the cost of reconnecting would be significantly higher because each bit of state must be read on each reconnection.

The alternative approach taken in Bluetooth low energy is that the server has to remember not only that a characteristic is configured for notification but also that the value has changed while a client is disconnected. This requires an additional bit to be associated with each notifiable characteristic that would be set when the characteristic value changes. Then, when the client reconnects, these bits can be checked and the characteristic value notified only if it did change.

The consequence of the server saving the configuration information for a bonded connection is that the client doesn't "connect" with a server. Instead, the client bonds and configures a peripheral to perform a function and then the client disconnects and reconnects to the peripheral, as and when it needs. The peripherals will be connectable when they have data to send, and the central device will reconnect to these bonded devices and quickly receive the data, as needed.

This is even true if the peripheral is doing something that the central device requested. For example, a central might ask a peripheral that has a time synchronization service to start synchronizing time and then disconnect. Some time later, the central device can reconnect to the peripheral and check on the state of the time synchronization, albeit with no guarantee that the synchronization completed.

The other major advantage of this approach is that the client can remember information about the handles of attributes in the peripheral. This means that once the central device has determined the set of services on a peripheral and configured them, the client can remember the attribute handles. When the central device reconnects to the peripheral, it can read and write these attribute handles without doing another service scan. This reduces the time required between reconnecting to a device and being able to use this device.

13.8 Changed Services

As stated in the preceding section, a central device's client can remember or cache the sets of services and characteristics between connections. Some devices will have the capability to change or add services. For example, a computer or a smart phone can add an application that might include a service, or a peripheral might have updated its firmware. When this happens, the client will not be able to read any attributes, and all requests to that server will fail. This protects the client from reading the wrong attributes.

Along with these error messages, the client will also receive a notification of the Service Changed characteristic from the GATT Service. The client should have stored this attribute's handle, so when the client receives this notification, it can take the appropriate action.

The Service Changed notification includes a range of handles that have been changed. This means that if a device has only added or removed one service, only the range of handles for that service will be included in the notification. If the device has had all of its services changed—for example, if the operating system on the device has been updated—the handle range might include all the handles of the device.

Note that Service Changed is only relevant for bonded devices. If the central device does not bond with a peripheral device, then no attribute handles can be cached by the client and no notification of any service changes will be made. This means that for two devices that are not bonded, the client must refresh the list of all services and characteristics on the server every time they connect.

13.9 Implementing Profiles

When designing a central device, the biggest implication is what it supports. Usually, a central device will implement the client roles of one or more profiles. To understand what this means, profiles must be explained.

13.9.1 Defining a Profile

A profile is a description of how a device functions for a given use case. Within a profile, roles are defined. A profile role defines a device that can act as one part of an ecosystem of devices that enable the profile's use case.

Each profile role defines the set of services that a device must implement. Some profile roles don't require any services, others require just one service, and others require many services. Typically, profiles define two roles, for example, a reporter role and a monitor role. The reporter role would require one service and would be implemented in a peripheral device. A monitor role would not require any services and would be implemented in a central device.

For example, the proximity profile defines two profile roles: the proximity monitor and the proximity reporter. The proximity reporter implements the Immediate Alert

Service, the Link Loss Service, and the Tx Power Service. These three services act independently on the reporter, behaving as defined in each of their associated service specifications. This independence of services is critical to this profile model being flexible and therefore future proof.

The proximity monitor defines how a client uses these three services to enable the proximity use case. For example, once the services and characteristics have been discovered, the monitor can read the current Tx Power characteristic and compare it with the received signal strength of packets from the reporter. The proximity monitor can then ensure that the implied distance between the two devices is acceptable. If this distance is not acceptable, the proximity monitor can write the alert level characteristic in the Immediate Alert Service.

The proximity monitor can also write the alert level characteristic in the Link Loss Service. In this way, if the signal strength drops so quickly that the connection drops before the signal strength drop is noticed or the alert level characteristic in the Immediate Alert Service can be written, the proximity reporter can still alert in a controlled way.

This combination of services allows the proximity monitoring use case. But the Immediate Alert Service could also be used to enable different use cases, such as allowing the user to see which device is selected on a user interface by writing the alert level characteristic of the Immediate Alert Service. This behavior in the client would be described in a different profile.

13.9.2 Finding Services

The first thing that a profile needs to do is find the services that the peer device supports. It can do this by first discovering primary services. This can be done by either finding primary services by service type or by finding all primary services.

This makes it possible for a profile that only requires the use of a single service to just find the particular service it needs, ignoring all other services on the device. For example, the user interface alerting service would only need to find the single service that it uses.

A more complex profile might require that many services are found. The proximity monitor profile role requires that three services are discovered. It could do this by doing the same single service search, one at a time, or it could do this by finding all primary services and then only storing the services that it needs.

Typically, a complex client only finds all services once and then caches them for later. When an application that implements a given profile role asks for all the services on a device, the cached list of services is used. This is possible because when the set of services changes on a server, a notification is sent to inform the client that the services have changed.

13.9.3 Finding Characteristics

Once the set of services has been found on the peer device, the set of characteristics will be required. Just like services, only one characteristic might be required by the profile, or many characteristics might be required. Some characteristics might be optional, and as such, they must be searched for within a service to check whether they exist.

Typically, all characteristics for each service previously discovered will be found by the client. This is because for the simplest service that has just one characteristic, this is a very fast and efficient operation. Also, if the service contains many optional characteristics, this operation finds all the characteristics the service implements.

13.9.4 Using Characteristics

Once the characteristics have been found, they can be used. For example, for the Immediate Alert Service, the proximity monitor client can write the required alert level into the Alert Level characteristic value.

It should be noted that for some services that have just one characteristic that is readable, the characteristic discovery and the reading of the characteristic value can be combined into a single operation. For example, once the Tx Power Service has been discovered, the Tx Power characteristic can be discovered and read by using the Read Characteristic By Type GATT procedure. This means that the separate step of finding the characteristics that was just described is not required.

Some characteristics can support being notified or indicated. For a client to be able to use these capabilities of a server, the client must first find the Client Characteristic Configuration Descriptor. The client does this by finding any additional attributes after the characteristic value, within this characteristic group. This will find all additional descriptors for the characteristic. Once the Client Characteristic Configuration Descriptor has been discovered, it can be written with the correct value to enable notifications or indications.

13.9.5 Profile Security

Finally, if the client wants to disconnect and reconnect quickly again in the future, or the characteristics require an encrypted or authenticated link, the client must bond with the server. Typically, this is driven by the client role in the profile.

The client can attempt to read or write any characteristic value in a service. The service only has to respond with the value of the characteristic or with the response that the value was written, if the correct permissions are in place to read or write this characteristic. If the permissions are not correct, the server will respond with an appropriate error code.

13.9 Implementing Profiles

If the error code suggested that the client had insufficient security, the client could attempt to pair or bond with the server to enable the correct level of security. If the client is just reading the information once, pairing would be sufficient. If the client will reconnect again and again, bonding would also be required. Once the devices have the appropriate security, the connection will be encrypted, and the client can retry the request that previously failed.

Chapter 14

Peripherals

The hand is the cutting edge of the mind.
—*Jacob Bronowski*

This chapter looks at the design of peripherals, tying together all the parts of this book from the perspective of a peripheral.

14.1 Background

Peripherals are the lifeblood of the Bluetooth low energy ecosystem. While a central device will typically be in a phone, television, or computer, peripherals are custom-designed products that are heavily optimized for ultra-low power consumption.

Peripherals are mostly designed around their battery, their sensors, their inputs and outputs; only then is the wireless technology considered. For peripherals to work, they must interact with central devices. They can do this in three different ways: they can broadcast data; they can be discovered and connected to by a central device; or they can stay disconnected and then establish connections as they need.

14.2 Broadcast Only

Some devices will have a single quantum of data that they want to share with many other devices. The best and lowest power way to do this is by using the broadcasting model. For example, a device that knows the current time and wants to share it with every other device in the area would broadcast that time data so that other devices can receive the current time.

A peripheral that only does broadcasting can be very power efficient. It does not need to be discoverable or connectable. It does not need to accept any connections from central devices or have an extensive attribute database that can be discovered. It just constantly broadcasts useful data.

One consideration is the broadcast interval or frequency at which the device will broadcast. For example, it is possible to broadcast the current time every 100 milliseconds, or 10 times a second. But given that the current time is typically only accurate to the nearest second, it is not worth wasting the energy broadcasting the time repeatedly within a one-second interval. Therefore, the interval at which some data is broadcast may be based on the time that a user is willing to wait before the data is available.

The current time is probably something that can be broadcast at reasonably slow intervals because the observers of this information will only listen infrequently for this information. Say, once a day, or even once a week. Thus, a broadcast interval of once every 10 seconds would then be reasonable.

It is interesting to note that a wall clock that synchronizes its time from a broadcaster once a day only needs to be accurate to the nearest second over the course of that day. A wall clock that would normally be off by half a minute a month would be made more accurate by synchronizing with an accurate time broadcaster once a day.

14.3 Being Discoverable

Apart from peripherals that only broadcast, all other peripherals will start off being discoverable. Discoverable means that the peripheral is advertising to any scanning central device in the area that is looking for peripherals.

There are two types of discoverability: limited and general discoverability. A peripheral is only limited-discoverable for a short time after interacting with the user. At all other times that the peripheral needs to be discoverable it would be generally discoverable.

For example, when the batteries of a peripheral are first installed by the user, it would be limited-discoverable. This makes it possible for a central device, probably held by the same user, to display this new device at the top of a list on the user interface.

Peripherals would also be limited-discoverable immediately after a connect button is pressed on the device. Again, the user would expect to see the peripheral at the top of the list of devices after he has pressed that connect button.

Peripherals don't need to be discoverable all the time. It is very common for devices that have paired with a central device to never be discoverable again or to only become discoverable again after the connect button is pressed. This has two advantages.

First, peripherals that are not discoverable don't need to include information in their advertising packets that is related to discoverability such as the device name or the current Transmit Power.

Second, central devices that are looking for devices to discover will only want to display devices that are of interest to the user, and not all peripherals in the area. By peripherals not being discoverable all the time, the list presented to the users of central devices will be much more manageable and useful.

14.4 Being Connectable

Being discoverable and connectable are very similar to one another; they both use advertising packets, yet they serve different purposes. Discoverable devices are typically not paired with any other device, whereas connectable devices are typically bonded with one or more devices and would only accept connections from those devices.

When a device is connectable, it accepts connection requests from initiating central devices that want to connect to it. Essentially, when a peripheral is connectable, it accepts connections from any device that sends it a connection request packet.

This promiscuous behavior is not great from a power consumption point of view. If any central device can connect with the peripheral, it can take lots of time talking with the peripheral, wasting the peripheral's power, and probably more important, preventing the peripheral's bonded centrals from connecting to it.

To solve the this problem, the controller in a peripheral device can be configured to only accept connections from a limited set of central devices. The set of central devices is stored in a white list, and any connection request from a device that is not in that white list is ignored. This way, the peripheral need only accept connections from the central devices that are bonded and in the white list; thus, the peripheral only uses power to talk with its bonded devices.

This model of only accepting connections from bonded devices in a white list can only be used once the device is bonded. Before this is enabled, the device must be promiscuous and accept connections from any device. Typically, this is only when the device is initially connectable and not bonded with any devices yet.

14.5 Exposing Services

Once in a connection, the peripheral typically exposes one or more services. Each of these services encapsulates the atomic behavior of a component of the peripheral. These services are exposed through attributes by using an Attribute Protocol server. The collection of attributes in a server is normally referred to as an attribute database.

The organization of these attributes is determined by the Generic Attribute Profile (GAP). It has the following structure:

Each service in the attribute database starts with a service declaration that defines the type of the service. The attributes belonging to the service follow this declaration. Next comes the service declaration for the ensuing service.

Within each service, there can be one or more other services that are included by a service. Included services make it possible to incorporate more complicated behavior within a service without extending the core functionality of that service. This encourages the definition of services that only implement some small piece of functionality.

For example, the Battery Service only exposes the battery level. A device like a camera might have two batteries, one for the flash and one for the main camera body. These components can have their own services; for example, the Camera Flash Service and the Camera Service. The Camera Service would therefore include both the Battery Service for the main camera and the flash service. The Camera Flash Service would include a separate Battery Service for the flash.

Also within each service, there can be one or more characteristics. A characteristic is fundamentally a single value that can be accessed. But each characteristic also has a characteristic declaration that defines what the data is, how it can be accessed, and can also include additional information that describes how it is formatted or how it is configured.

Each of these service declarations, service includes, characteristic declarations, characteristic values, and characteristic descriptors are individual attributes within the attribute database. Therefore, the peripheral just exposes the attributes it needs to expose, depending on what behavior it is exposing, and then waits for a client to come along and interact with these attributes.

14.6 Characteristics

Characteristics are the fundamental building blocks of services. A characteristic is just a value with which a client can interact. The format of the value of a characteristic is determined by the characteristic type, and ultimately by an XML file that is the characteristic specification.

The behavior of a characteristic is not defined by this XML file; rather, it is defined in a service specification. Therefore, within a single peripheral, it is possible to have multiple characteristics with the same type, but different behaviors.

For example, the Alert Level characteristic is used in both the Immediate Alert Service and the Link Loss Service, but the behavior is different. In the Immediate Alert Service, the Alert Level characteristic is only writable and causes the peripheral to make an immediate alert based on the value written. This value is not readable because the value of the characteristic is consumed immediately; this is known as a *control point characteristic*.

In the Link Loss Service, the Alert Level characteristic is both readable and writable and causes an alert only when the central disconnects from the peripheral, based on the value of the characteristic at the point of disconnection. In this service, the characteristic has state and can be read and written. When it is written, the value is stored and can be read at a later point in time.

14.7 Security Matters

A peripheral device can expose data that could be considered confidential or private, or should only be sent to authenticated devices. To ensure this, the attribute server refuses any request for information about an attribute that it considers cannot be completed within the current level of security.

For example, if the client requests to read the number of unread emails from a phone or wants to write the ringer level of a phone, these should only be acceptable from a device that has been authenticated and for which confidentiality can be assured. This requires that the peripheral first authenticates and then bonds with the requesting central device. Typically, it is the central device that initiates the authentication and bonding. Once the devices are authenticated, the connection can be encrypted and the client requests can be re-sent.

This security model is very simple for a peripheral. For each request received by the peripheral's attribute server, the server first checks the security permissions required to accept the request. If the security permissions are insufficient, the server will send back an error response with a suitable error, giving a hint to the server about how it can go about fixing the problem before resending the request. If the security permissions are acceptable, the server will act upon the request and send back a response. At no point does the peripheral have to initiate any security requests.

The security that is required by a service or a characteristic within a service is defined by the service specification that is being implemented. Some services might require no security because they are exposing public information such as the current time. Some services might require very strong security, particularly if they are exposing private personal information.

14.8 Optimizing for Low Power

For peripherals to be able to operate with only the power provided by tiny batteries, for extended periods of time, consideration must be given to optimizing the peripheral for low-power operation. This includes determining the best advertising and connection intervals, optimizing access to attributes, and choosing whether to stay connected or to disconnect and reconnect.

To understand how low power can be achieved, consider the typical states that a peripheral device uses (see Figure 14–1). This starts with the peripheral being off. Obviously, when off, it is using no (or very little) power. When the device is first powered on, it moves to the discoverable advertising state. In this state, the peripheral is discoverable and can be found by one or more central devices.

At some point, a central device will connect with the peripheral and then bond. Upon bonding, the peripheral moves to the connectable advertising state. If bonding

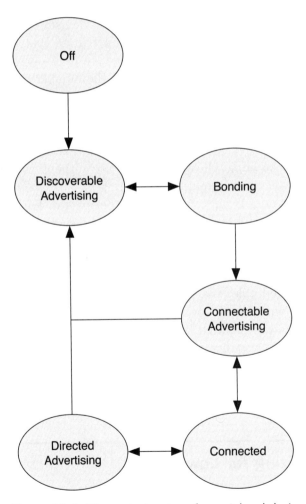

Figure 14–1 The typical states of a peripheral device

doesn't happen, the peripheral moves back to discoverable advertising and waits for another central device with which to connect and bond.

Once the devices are bonded, the peripheral will use connectable advertising. Then, only central devices that have paired with this peripheral can make a connection to the peripheral. Once they do reconnect, they move to the connected state and manage their connection intervals, depending on what the central device is doing.

When connected, the peripheral also needs to determine if it should disconnect now, or stay connected. This decision is more relevant if the central device has not made any requests recently, or if the peripheral has not sent any data to the central device or is not expecting to do so for some time.

When it does disconnect, it can move to one of two possible states. One would use the connectable advertising, in which the peripheral is periodically advertising so that the central device can reconnect as it needs. The other option is that the peripheral doesn't advertise at all until it has some data to send, at which point it would use directed advertising to reconnect to the central.

Finally, don't forget that the peripheral might have a connect button that when pressed removes the current bond and allows the peripheral to be discoverable again. Any devices that were connected at that point would be disconnected, and the device returns to the discoverable advertising state.

14.8.1 Discoverable Advertising

One of the most fundamental ways to optimize for low power in a peripheral is to choose appropriate intervals for advertising and connection intervals. The choice of good values could mean the difference between a device that has a few weeks' battery life and one that has a few years' battery life. But the choice is not simple because there are many compromises that must be made.

A peripheral device will typically start out performing discoverable advertising so that central devices can find it. The time that the peripheral is in this mode of operation should typically be very short in the total lifetime of the device because the user will want to take the device out and connect to it as quickly as possible. Once connected, the device will revert to only connectable advertising.

When in this discoverable mode, the peripheral will want to be found quickly, to provide the best user experience possible. It will also be advertising lots of additional information such as the transmit power so that the central device can sort devices by proximity; the device name to build a good user interface; and the set of services that the device supports to enable the central device to filter devices if necessary.

A discoverable peripheral device should therefore be willing to advertise at a fairly rapid rate to be discovered quickly. An advertising interval of about 250 milliseconds would be a good compromise between the speed of discovery and power savings.

14.8.2 Bonding

Whenever the peripheral engages in a connection, the connection intervals are determined by the central device. The central device should use sensible connection interval values—but this is not guaranteed.

Once the peripheral makes a connection from the discoverable advertising state, the connection interval can be fairly fast. A fast connection interval—for example, between 7.5 milliseconds and 25 milliseconds—can use lots of power, but it also means that the central device can discover the set of services and characteristics that this peripheral offers, and therefore can provide prompt feedback to the user about how it can interact with the peripheral.

If the connection interval is very slow (between 1 second and 4 seconds), it can take an extremely long time before the central device determines how to utilize the peripheral and the user might think that the device is not working.

Once the device is bonded and the central device has finished analyzing the device, it should be able to reduce the connection interval significantly to save power. The correct interval to choose is discussed a little later on. The central device can also start using the exposed services on this peripheral. At some point the device will disconnect.

Once the bonded peripheral is disconnected, it could move to the connectable advertising state, or to the directed advertising state. Each of these options is detailed in the following sections.

14.8.3 Connectable Advertising

When the peripheral is disconnected, it can periodically advertise to allow the central device to reconnect. The advertising interval used in this state is a compromise between how fast a central device can reconnect to this peripheral and the power consumption that the peripheral will use when disconnected.

For some peripherals, the time when not in a connection will be significant. For example, a heart-rate belt might only be connected while the user is running for an hour, three times a week. Each week, the heart-rate belt is therefore connected for three hours, but for the other 165 hours, it is in a connectable advertising state. Therefore, the peripheral should seriously consider the benefits of using a longer connectable advertising interval.

A connectable advertising interval of 1 second would allow a central device to be able to connect within a few seconds. This is probably a perfectly acceptable time for the average user. However, if the device needs to allow the central to make a connection quicker, then it would have to decrease the connectable advertising interval to 0.5 second or less.

It should also be noted that some devices would not need to advertise continuously. Again, let's look at the example of the heart-rate belt. When this peripheral is not being worn, the device does not need to use connectable advertising. It can therefore disable advertising when it's not being worn and save a significant amount of power when it is disconnected.

If the heart-rate belt is worn, it can then use more power when it is advertising, perhaps using a 100 millisecond advertising interval. This gives the user the impression that once she puts the belt on, it connects instantly—an excellent user experience. Obviously, if the central is not ready to initiate a connection with the heart-rate belt at this time, it would be prudent for the heart rate belt to move to a slower advertising rate, using a longer advertising interval, to save power in those cases when the belt is being worn but the central device is not within range.

14.8.4 Directed Advertising

Some peripherals will want to connect directly with a central device when something happens. For peripherals, where the time between something happening and when it sends a notification about the event to the central device must be as short as possible, it is best to use directed advertising.

Directed advertising burns a lot of power on the peripheral because the peripheral transmits lots of advertising packets very quickly to a single central device. If that central device is ready to initiate a connection to this peripheral, it will immediately connect, allowing the peripheral to send whatever data it needs, quickly.

Directed advertising is also the quickest way for a peripheral to make a connection to a central device. Connection times of fewer than 3 milliseconds are possible, including the transmission of application data.

There is no way to configure the intervals used when using directed advertising because the advertising packets must be sent every 3.75 milliseconds on each of the 3 advertising channels. This means that there will be one directed advertising packet transmitted by the peripheral every 1.25 milliseconds, or 800 packets per second.

For a peripheral that must send data very quickly, and that also rarely needs to be connectable, directed advertising is probably the best model.

14.8.5 Connected

When in a connection, the central device has complete control over the connection intervals and latency used by the peripheral. The peripheral does have the option to make a request to the central's host, by using a Logical Link Control and Adaptation Protocol (L2CAP) signaling channel command to suggest to the central device that the values currently being used are not useful.

For example, if the central device is monitoring the proximity of the peripheral device it would only need a connection interval rate of perhaps three times a second. If the actual connection interval used is shorter than this, the peripheral would be forced to use too much power synchronizing with the central with no user benefit. If the connection interval used is longer than this, the peripheral would possibly be synchronizing so slowly that the pair of devices would not be able to detect the movement of the devices early enough to warn the user.

There are two configurable values for the connection parameters that relate to power consumption: the connection interval and the slave latency. The connection interval is a time that determines how often the central will transmit and synchronize with the peripheral device. This is any multiple of 1.25 milliseconds.

The connection interval is not the most significant factor. The slave latency is much more important from a peripheral power consumption point of view. The slave latency determines the number of master connection intervals that the slave can ignore. This is a value from 0 to 500.

For example, suppose that the connection interval is 12.5 milliseconds and the slave latency is 0, then the slave would have to listen every 12.5 milliseconds for the master. This would burn a lot of power. With the same 12.5 milliseconds connection interval but a slave latency of 1, the slave can ignore one connection interval but must listen to the next one. This halves the power consumption of the slave, yet allows the slave to have the same ability to send data within 12.5 milliseconds if necessary.

The slave latency is not just a way for the slave device to save power; it also determines the latency that any data from the master can be sent to the slave. For example, if you have a keyboard with a Caps-Lock light, then this light might need to be turned on and off within 0.5 second. If the connection latency was 12.5 milliseconds, slave latency would have to be 39. The slave could then skip 39 out of every 40 packets from the master, and therefore would listen once every 500 milliseconds. This would also allow the slave to save a significant amount of power compared with a slave latency of 0 or 1. Of course, this is an approximation because it assumes a perfect channel without the need for retransmissions. In practice, retransmissions will be required, so sometimes latency will be greater.

It should be noted that this is not a never-ending benefit. For example, if the data flow were only from the peripheral to the central device, it could be possible to use a very high slave latency. However, using a very high slave latency yields diminishing returns.

For example, consider a mouse that only sends data to a computer. If the connection interval is 15 milliseconds, and the slave latency is set to 500, then the mouse would only need to listen once every 7.5 seconds. But consider the fact that both the central and the peripheral are timed based on clocks that have a certain level of

14.8 Optimizing for Low Power

inaccuracy. In the worst case, these inaccuracies can amount to 500 parts per million on each device. In other words, every second, the timing of either the central or the peripheral might be off by up to half a millisecond.

This means that after the 7.5 seconds of time that the mouse has not been in synchronization, its concept of time could be off by 3.75 milliseconds, as would that of the master, possibly in different directions. Therefore, the total uncertainty would be 7.5 milliseconds. The peripheral has to listen for an additional 7.5 milliseconds before the time when it has calculated the master should be transmitting, and up to 7.5 milliseconds after. This is called *window widening*. This burns power. So every 7.5 seconds, the radio has to be on for 7.5 milliseconds (best case).

If the slave latency was only 50, the mouse would have to listen once every 750 milliseconds, but the window widening would mean that the slave has to listen 0.75 milliseconds early. Unsurprisingly, this requires the same proportion of time for window widening as the previous example.

The peripheral also must receive and transmit some empty packets, an 80 microsecond packet from the master and an 80 microsecond packet in reply. As the slave latency increases, the time taken to transmit and receive the packets becomes insignificant; therefore, the additional power savings from not having the slave synchronize frequently is lost to window widening.

As a result, it makes no sense to set a slave latency of greater than about 1 second or fewer than 300 milliseconds. Below this range, the power used to repeatedly synchronize is higher than it would be to wait a little bit longer. Above it, the power used by window widening doesn't save any significant amount of power, and so it is better to use the lower latency to enhance the user experience.

14.8.6 Stay Connected or Disconnect

When the peripheral is connected, the use case might require that data is transferred at a random interval. A peripheral that is performing in such a use case could stay connected or disconnect and then reconnect when it has some more data.

There are two main decisions to make. First, can the central device reconnect back to the peripheral in a reasonable latency if the peripheral starts advertising so that the users don't know that it was disconnected? Second, if it does stay connected, what connection latency is being used or is possible to ask for, such that the battery life is still acceptable?

To enable the peripheral to make more informed decisions, the Scan Parameters Service can be exposed on a peripheral so that a central device can inform the peripheral what scan parameters it is using. When the central device connects to the peripheral, it discovers this service on the peripheral, and the central device can then write the latency that it will honor to the peripheral when reestablishing a connection. Using this information, a peripheral can determine if and when it should disconnect, given the excepted user experience.

For example, if the connection latency written by the central device into the peripheral's scan parameters service is 100 milliseconds, and the peripheral is happy with a latency of 250 milliseconds, then it could theoretically disconnect at any time if it has no data to send. The device knows that it can reconnect within 100 milliseconds and be able to send data. However, although disconnecting saves power, the reconnection procedure uses lots of power. With a 100-millisecond reconnection latency, the peripheral will spend on average 50 milliseconds to make a connection. During these 50 milliseconds, the peripheral will be sending a 176-microsecond packet every 1.25 milliseconds, therefore powering the radio for 8.64 milliseconds.

This doesn't sound like a very long time. But if the peripheral were connected with a connection interval of 250 milliseconds, and it was not sending any data, it would only have the radio on for 640 microseconds every second. Therefore, the peripheral would have to be connected to the central device for 13.5 seconds before it would use more power staying connected than disconnecting and then reconnecting.

However, unfortunately, it's not that simple. If the peripheral also wants to be connectable, then it would also need to advertise slowly when disconnected. Suppose that the peripheral has a 1-second advertising interval. It would have to send up to 3 advertising packets that take 504 microseconds. So, the peripheral could stay connected and use 640 microseconds of time every second to stay connected, or it could disconnect, advertise periodically, and then rapidly advertise when it needs to make a reconnection.

Therefore, the peripheral is only saving 136 microseconds per second by disconnecting. This means that the peripheral would actually need to stay connected for over a minute before it makes sense to disconnect. It gets worse: If the peripheral wants to allow a central device to connect quicker, it would need to advertise quicker. An advertising interval of 500 milliseconds would actually use more power than just being in a connection. Therefore, for some devices, it actually makes sense to stay connected all the time.

This implies that for some peripherals, there is no need to implement this service. Once the peripheral is connected, it would remain connected. This really depends on the use case and how the peripheral is used.

Of course, if a peripheral has exposed the scan parameters service but the central device doesn't write any useful values into the service characteristics, then the peripheral must use guesswork to determine when it should stay connected or it should disconnect. This guesswork is typically a simple matter of staying connected for a given period of time after the last application data was sent. Once this time-out has expired, the peripheral would disconnect. Typical values for this might be measured in the order of minutes or even hours.

14.9 Optimizing Attributes

At the end of the day, the typical use of a peripheral is to provide access to the data it generates. For example, a heart monitor exposes the heartbeat rate of the person who is wearing it. The final optimization that a peripheral can use is how this data is transferred.

Typically, a peripheral exposes one or more services. The central device discovers these services and their characteristics and descriptors. The central device then reads, writes, and configures these characteristics to utilize the services offered by the peripheral. But the peripheral can help itself to save power by implementing notifications and indications about its characteristics.

Characteristics and services are described by using attributes, and these are accessed by using the attribute protocol. This protocol enables a client (in a central device) to access these attributes in a peripheral. The protocol not only makes it possible for attributes to be read and written but also notified and indicated.

Take a characteristic of a device—for example, a measured heart rate—that is being transmitted once per second from the heart-rate monitor to a central device such as a watch or phone. This data transmission could be implemented by reading the characteristic's value once every second. But this would require a read request to be sent by the central and a read response sent in reply by the peripheral. The peripheral's radio would be active for 272 microseconds every second just to send the request and receive the response.

If the characteristic is configured for notifications, this would reduce the active time to just 232 microseconds. This is significantly more energy efficient than the polling case. This advantage is more evident when the data is only available at random times and not periodic. In this case, the client might be polling for data more often than is necessary.

Consider a device that monitors a sensor once per second, but the value only changes infrequently, perhaps once every 30 seconds. In this case, polling by using a read request and read response would need to be performed 60 times per minute, but the value might only change once during this time. This requires 60 requests to be sent, and 60 responses to be received in return. This takes 16.32 milliseconds every minute. But, if the peripheral supports notifications, and the central device configures notifications, the radio would only need to be active for 9.744 milliseconds. This is almost half the power consumption of the polling case.

Indications are similar to notifications, except that they also have a response message to acknowledge at the protocol layer that the information has been received. This uses slightly more energy than the preceding notification case, approximately 10 milliseconds, but this has the advantage that no data is lost. Because notifications are not acknowledged at the protocol layer, they can be ignored if the receive buffers are full. But indications can only be sent one at a time and cannot be sent again until a confirmation message has been received.

Notifications and indications are therefore highly efficient when compared with a polling model. A peripheral can help itself be more energy efficient by ensuring that as many characteristics as possible are exposed for notification or indication. This makes it possible for a central device to configure characteristics to be notified or indicated.

Chapter 15
Testing and Qualification

Ideas must be put to the test. That's why we make things; otherwise, they would no more than ideas. There is often a huge difference between an idea and its realisation. I've had what I thought were great ideas that just didn't work.
—Andy Goldsworthy

Before releasing a Bluetooth low energy product that implements one or more services and profiles, it must be tested and then qualified. The Bluetooth Special Interest Group (SIG) requires this and codifies it in the membership agreements that all Bluetooth SIG members sign. This agreement allows products that comply with the published specifications to be able to utilize and rely on the necessary claims of the intellectual property that any other members may have. This means that a product that has implemented one or more Bluetooth specifications, as determined by a Declaration of Compliance (DoC) signed by that member, has a royalty-free license to use that technology. This implies that there is a great benefit for declaring compliance. Proving that your device is compliant is where the testing and qualification come into the picture.

As illustrated in Figure 15–1, the qualification program is a multistep process that spans product conception through to a qualified and listed product. Part of this process requires that testing is performed. Some steps require that a monetary investment is made; others require decisions to be made about what the product will do.

15.1 Starting a Project

The very first step is the easiest. You start a project. This requires you to log in to the Bluetooth SIG Web site[1] at bluetooth.org and start a new Test Plan Generator (TPG) project. You need a project name and the date that you expect to qualify. Don't worry about this information too much. You can change these names and dates in a later step.

---

1. To create a new project, go to http://www.bluetooth.org/tpg/create_project.cfm

314 Chapter 15 Testing and Qualification

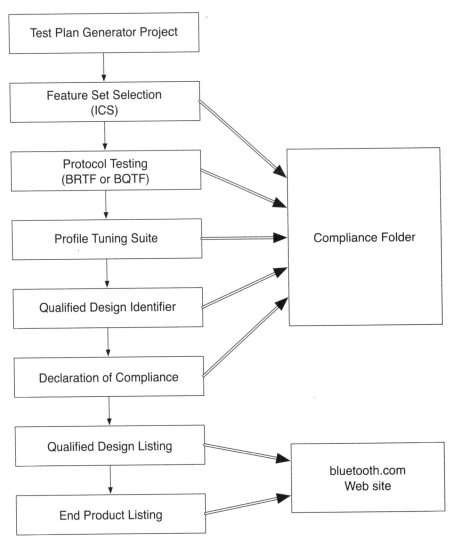

Figure 15–1 The Bluetooth testing and qualification process

The nice thing about starting a project is that it is free. You can start as many projects as you want. If you want to test out what might be required to do something, you can just start a project, configure the project, and see what the resulting testing burden is. You can also delete projects at any time, so if a project no longer looks viable, you can remove it from your current list of active projects.

The project that you create in this initial step will be the same all the way through to the point of qualification. This means that although the project is free to create, before the project can become an officially listed, qualified product, a fee will have to be paid.

15.1 Starting a Project

To start the project, you go to the Create New Project page on the bluetooth.org Web site. The information you will need includes the expected qualification date, the TPG Release Version, a project name, product type, and a description of the hardware and software versions.

For a Bluetooth low energy product, the TPG Release Version should be 4.0 or later. By doing so, the project takes advantage of all the low energy layers in the system that were only introduced in the v4.0 core specification.

The next value to select is the product type. There are eight types of products that can be selected, depending on the type of component or subsystem that is being qualified.

- **End Product** This is an actual product that is available for public consumption.

- **Host Subsystem Product** This is a product that is only composed of the host parts of the core specification, such as Logical Link Control and Adaptation Protocol (L2CAP), Attribute Protocol (ATT), Generic Attribute Profile (GATT), Security Manager (SM), and Generic Access Profile (GAP). This needs to be combined with a controller subsystem and one or more profile subsystems to make an end product.

- **Controller Subsystem Product** This is a product that is only composed of the controller parts of the core specification, such as the radio and link layers. This needs to be combined with a host subsystem and one or more profile subsystems to make an end product.

- **Profile Subsystem Product** This is a profile subsystem that contains an implementation of one or more profiles and/or services. This will be combined with a host subsystem and a controller subsystem, and possibly additional profile subsystem products to make an end product.

- **Component Subsystem Product** This is an individual part of a subsystem. For example, the implementation of your SM might be separately listed as a component subsystem product.

- **Development Tool** This is a special product type that you use during the development of Bluetooth products. This type of product cannot be sold to the end consumer, but it can be distributed to other members to assist the rapid creation of end products.

- **Test Equipment** This is a special type of product designed for the small set of test equipment manufacturers who need to have access to the royalty-free license. Because the test equipment might need to do things outside the specification to enable testing of real products, these tools have a special product type.

The interesting thing about this set of product types is how they can be combined. At the most discrete level, everything can be combined together into an end product as a whole. This holistic approach can cause lots of problems because the complexity of the testing will be huge. Instead, most components would be tested individually and listed as individual component subsystems. These can then be combined together without any additional testing into an end product.

Components can also be combined into a controller or host subsystem. By doing this, whole parts of the product's design can be combined into a single listing. For example, when using a controller chip, it would be listed as a controller subsystem so that it can be combined with your host subsystem to make an end product.

15.2 Selecting Features

The next step in the process of qualifying a product is to select what features your product is going to support. This is declared in the summary template. Here, you can select the set of core layers and profiles and services that your device will support. For example, a low energy proximity device would select RF PHY, Link Layer, L2CAP, ATT, SM, and GATT in the core section, and Link Loss Service, Immediate Alert Service, and the Tx Power Service in the host section.

After this, you need to delve deeper into the details of what is being claimed. This is where you match the features that you plan to implement with the feature set for which you will later be claiming compliance. For every feature claimed, a set of tests will be required to be run, and test evidence will need to be collected before the DoC can be signed.

For each layer that was selected in the summary template, the detailed Implementation Conformance Statement (ICS) is filled in. For each ICS, each and every possible mandatory or optional feature has a line where you can identify whether your product will support this feature. There is an ICS for each layer of the core specification and each profile or service being claimed.

15.3 Consistency Check

Once all the ICSs have been filled in, the validity of the selections can be checked. There are many rules as to what features you must have if another feature has been selected. If any of these selections are invalid, the line will be marked as invalid. Once these errors have been corrected, the consistency check can be refreshed. For example, if a service has optional notification of a characteristic, then if you select this line, the consistency check will ensure that GATT also has the ability to write characteristic descriptors. This is because without the ability to write characteristic descriptors, it would be impossible to configure a characteristic to enable notifications.

Sometimes, the set of features selected will be inconsistent. If they are inconsistent, the set of features selected will need to be changed and the consistency check attempted again. Once all the consistency checks have completed, it is possible to move to the next step.

It is worth noting that occasionally the test plan generator (TPG) has errors for new specfications. If you believe you are correct and the TPG signals an inconsistency, this can be raised with the Bluetooth Qualification Administrator (BQA), who can arrange for the TPG issues to be fixed.

15.4 Generating a Test Plan

Each layer of the core specification, or service or profile specification, has an associated test specification. Each test within these test specifications has a unique identifier. For your product, many of these tests will not be required, depending on the configuration of your ICS that was determined from the previous step. Therefore, the next step is to generate your test plan for your product as it pertains to the ICS that you have just completed. The test plan includes all the tests that need to be run to prove that your product has implemented all the required functionality that you are claiming.

The test plan that is generated can be downloaded to your local computer from the Web site; this provides you with a list of tests to use during your product development. Once the product has been developed and a prototype is available for testing, there is one final step before the qualification testing can begin. This is the creation of a compliance folder.

15.5 Creating a Compliance Folder

A compliance folder is a single location where all the information about a product, that product's feature set and capabilities, and all test evidence are stored. The compliance folder can be a collection of electronic documents or paper documents. The folder must be stored securely and be easily accessible for each product that you have qualified. If later on a problem is found with your device, the Bluetooth SIG can audit that product's compliance; thus, it must be given access to the compliance folder to be able to perform this audit.

The compliance folder contains the following items:

- **Product Information** This includes information about the product, such as the product name, a description of the product, and hardware and software version information. It also includes, where relevant, the user manual or user guide, a block diagram and technical description of the product, and notes on how the Bluetooth component is integrated into the design of the product.

- **Testing Information** This includes all the information about how the device was tested, including the test plan that was generated, a test declaration that has the result of each test, test reports that are required for formal validated tests, and any extra information that might be required to perform the tests so that the testing can be replicated at a later date.

- **Design Information** This includes detailed design information that describes the design of the product where relevant to the Bluetooth functionality. This could include elements, such as the circuit schematics, printed circuit board layouts, component layouts, and a bill of materials that describes the physical external components and their tolerances. This could also include things such as the software design documentation and sufficient information to extract the precise software used during testing from a software version control system.

- **Implementation Conformance Statements** This includes the ICS documents for each layer in the system and each profile or service.

The compliance folder must be stored securely while the product is offered for sale or distribution. Only after the product has been removed from sale for over a year can the compliance folder be destroyed.

15.6 Qualification Testing

Once the prototype is ready for testing, a test plan has been generated, and a compliance folder has been created to store all the information about the product and its testing information, the testing can begin.

For some tests, there exists validated test equipment that can be used to run the tests. There are three types of validated testers: RF testers are used to run the physical layer qualification tests; protocol testers are used to run the protocol layer tests; and the Profile Tuning Suite (PTS) is used to run the profile tests.

Each test is assigned to one of four different categories:

Category A Cat A tests are the best-defined test cases and must be done by a Bluetooth Qualified Test Facility (BQTF) or a Bluetooth Recognized Test

Facility (BRTF). The tests that are run will generate a test report that will be stored in the compliance folder.

Category B Cat B tests are run by the product company according to the test requirements. The test setup, test evidence, and test results are stored in the compliance folder.

Category C Cat C tests are also run by the product company, but the documentation requirements are significantly reduced. Cat C tests have to be declared that they have been performed and the test result must be captured, but no test evidence is required.

Category D Cat D tests are optional. There is no requirement to run these tests and there is no need to document that they have been run, that the results have been captured, or to store any test evidence.

Category D tests might appear to be mostly pointless, but all tests can move between categories as the tests mature or problems are found. For example, a new specification might start with all tests being at Category C, but once the maturity increases, these can move to Category B. Category B tests might be implemented by a test-equipment manufacturer that validates these tests, and, therefore, these might move up to Category A.

A test might be moved to Category D when it has known unresolved issues, and thus, the SIG cannot reasonably require that it be run. However, you should keep in mind that when you qualify, you declare compliance to the specification. Test cases are merely a useful way to check that compliance. So, even if a test case is broken, products must still correctly comply with the feature it was designed to test.

It is possible to raise test case errata on all the tests in the qualification program if the test case or the expected outcomes are invalid. The core specifications, not the test specifications, are the determining factor in deciding whether something is a valid test. It is therefore possible to take a Category A test, prove that it does not implement what is required in the core specification, and still qualify your product because you can still provide the test evidence required. This doesn't happen very often, but it is a useful safety valve.

15.7 Qualify Your Design

Once the testing has been completed, the compliance folder should be full of test reports. The next step is to qualify your design. To do this, you need a Qualified Design Identifier (QD ID). This is a unique number that refers to your design. This does not refer to an end product; this will come later. Instead, you qualify a design that can be used in multiple end products. For example, you might design a product that can be sold under multiple brands, with each brand using its own packaging. The qualification of the design allows the parts of the product that use Bluetooth wireless technology to be used by many end products. It does cost money to qualify a design, but this design can be used in many end products with no additional qualification charges.

Once the QD ID has been obtained, the test declaration is uploaded to the Bluetooth SIG Web site. This test declaration includes all the test reports from the validated testers as well as any additional Category C or Category D tests that you performed directly. Also, the profile tuning suite reports should be uploaded separately. These are directly parsed and the set of profiles and services that your device will support is immediately verified.

15.8 Declaring Compliance

Once the Bluetooth SIG has all the information about the design—what it does, the tests that have been run, and the test results—the design can be qualified. The member company that has designed the product now completes a Declaration of Compliance (DoC). This is a legally binding document that states that the member declares that the design complies with the Bluetooth specifications referenced, that the product implements the reference design, and that the product meets the testing and documentation requirements of the defining process document. This whole process is defined by the Qualification Program Reference Document (PRD). Essentially, by signing this declaration of compliance, the member has said that it has done everything necessary according to the process.

Because the DoC is a legal document, it can only be signed by someone who has the authority to enter into a legal commitment for the company. This means an officer of the company or someone who has been delegated the authority to make legal commitments by an officer of the company must be the signer.

The DoC must include the QD ID, the product information, the feature set of the product, the Bluetooth specifications being referenced, and the version of the PRD and test specifications used, as well as the test case reference list used to determine the set of tests required and their categories.

Once this DoC has been signed by a member company, the product is covered under the royalty-free license, and the Bluetooth brand can used in association with this product.

All products that have been qualified must have the QD ID clearly marked on the device or in its documentation. This allows anybody to determine the set of features and, therefore, the requirements of the design.

15.9 Listing

Once a member company has a compliant product with a QD ID, a listing is next performed. This listing does not need to be made public immediately because this can have an impact on the marketing plan for the release of the product. But a product must be listed within three months. The product compliance is not considered to be effective until the product is listed; therefore, you must list your product before offering it for sale or distribution.

15.10 Combining Components

It is possible to combine multiple listings into a single product. For example, an end-product listing might use a controller chip from one manufacturer, a host stack from another, one service from another supplier, and another service from yet another. This product would, therefore, have five QD IDs: one for the controller, one for the host, two for the services, and one for all the QD IDs in combination.

It should also be noted that each of these qualified designs can themselves be a combination of multiple, qualified components. The host in the previous example might have a separate QD ID for each layer in the design: one for L2CAP, one for ATT, one for GATT, one for GAP, one for SM, and one for this combination of components as a host subsystem. This seemingly simple end product would therefore ultimately reference nine different QD IDs.

This tree of QD IDs is a very powerful concept; it allows packaged components or subsystems to be designed and qualified separately and then delivered as a complete design. These components or subsystems can be reused in many products, again and again, without paying any additional testing costs or listing fees. For devices that use many existing parts, the costs for creating a new product can be negligible.

Index

Numbers

2.4GHz ISM band
  Bluetooth low energy using, 4–5
  overview of, 54
  at Physical Layer, 29
  transmit power, 56–57
3-Wire UART, HCI physical interface, 132–134
24-bit CRC, Bluetooth low energy. *see* CRC (cyclic redundancy check)
32-bit MIC, Bluetooth low energy. *see* MIC (message integrity check)
128-bit UUIDs (Bluetooth Base UUIDs), 190–191
10101010 packet sequence, transmitter tests, 63–64
11110000 packet sequence, transmitter tests, 63–64

A

Abstract state, 182–183
Abstraction, service-oriented architecture, 23
Access address
  Link Layer connections, 95
  packet structure, 30–31, 80–81
  test packet format, 63
Access permissions, attribute database, 194
Acknowledgement
  of data packet, 101
  optimizing for low power, 127
Action, requesting for command packets, 136
Active scanning
  in device discovery procedure, 257, 283–285
  HCI, 152–153
  Link Layer state machine, 72
  overview of, 72
  receiving broadcast data, 93
Active state mode, 3-Wire UART, 63–64, 133
Adaptive frequency hopping
  Bluetooth low energy design, 8
  channel map, 97–98
  data channels used with, 30
  defined, 9
  Link Layer connection process, 93–94, 97–98, 111–112
  Link Layer robustness, 120–122
  managed by master, 14
  optimizations for low power, 127
  overview of, 88–89
ADV_DIRECT_IND advertising packets, 81–82, 266–267
ADV_IND advertising packets, 81–82, 267
ADV_NONCONN_IND advertising packets, 82, 266
ADV_SCAN_IND advertising packets, 82, 266
Advanced Encryption System. *see* AES (Advanced Encryption System)
Advertisers, defined, 14–15
Advertising
  access address, packet structure, 80–81
  broadcasting data with, 42–43
  data, 273
  events, 90–92
  formatting data when broadcasting, 263
  Host/Controller Interface, 148–150
  initial discovery using devices for, 256–257
  interval, 90
  presence detection using, 41–42
Advertising channels
  access addresses for, 80–81
  advertising packets as transmitted on, 76
  in connection state, 74
  finding devices with, 90
  in Link Layer, 30–31
  overview of, 84–87
  reducing number to reduce power consumption, 70
  in scanning state, 72
  used by devices in broadcast mode, 263
Advertising packets
  broadcasting data with, 93, 148–150
  finding devices, 90–92

323

Advertising packets (*continued*)
  GAP connection modes, 266–267
  GAP connection procedures, 268–269
  HCI connections to white lists, 155
  header contents, 81–82
  length field, 83
  overview of, 76
  peripheral connectability, 300–301
Advertising state
  entering connection state from, 73
    entering slave substate from, 74
    nonconnectable advertising device in, 92
    optimizing peripherals for low power, 304–306
  overview of, 71
AES (Advanced Encryption System)
  calculating MIC, 107–109
  HCI controller setup, 145–146
  overview of, 105–106
  security features, 244
  starting encryption for connections, 114
AFH. *see* adaptive frequency hopping
Alert Level characteristic, 288–290
Algorithms, scheduling, 75
Alternate MAC PHY (AMP), Bluetooth version 3.0, 3
AM (amplitude modulation) radio, 50–51
AMP (Alternate MAC PHY), Bluetooth version 3.0, 3
Amplitude modulation (AM) radio, 50–51
Amplitude-shift keying (ASK), digital modulation, 52
Analog modulation, 49–51
Appearance characteristic, GAP Service, 276–277, 284
Application data rate, radio systems, 51
Application Errors response, 231
Application layer architecture
  characteristics, 36–37
  defined, 36
  profiles, 37–38
  services, 37
  three-chip solution, 39–40
  two-chip solution, 39–40
Architectural paradigms, concepts, 20–25
Architecture
  application layer, 36–38
  Bluetooth, 27–28
  Bluetooth low energy design as, 9
  controller, 27–31

  host, 32–36
  stack splits, 38–40
ASK (amplitude-shift keying), digital modulation, 52
Assembly, by multiplexing layers, 170
Asymmetric design concept, 14–15
ATM networks, as multiplexing layers, 170
Atomic operations and transactions, 197–198
Atomic services, 34
Attribute database
  accessing attributes, 196–197
  exposing services to peripherals, 301–302
  overview of, 192–193
  permissions, 194–195
Attribute handles
  Find By Type Value Request/response, 222–223
  Find Information Request/response, 221–222
  Invalid Handle error, 228–229
  overview of, 189–190
  Read By Type Request/response, 223
  Read Request including, 224
Attribute Not Found error, 230
Attribute Not Long error, 230
Attribute Profile, 199
Attribute Protocol
  attribute client using, 192
  Bluetooth low memory using only, 14
  channel identifier for, 172
  control points, 183
  creation of, 179
  error responses, 228–231
  Exchange MTU Request, 221
  exposing state with, 16–17
  Find By Type Value Request, 222–223
  Find Information Request, 221–222
  Generic Attribute Profile vs., 231
  Handle Value Indication, 228
  Handle Value Notification, 227–228
  host architecture, 33–34
  overview of, 217–219
  Prepare Write Request and Execute Write Request, 226–227
  protocol messages, 219–220
  Read Blob Request, 224
  Read By Group Type Request, 225
  Read By Type Request, 223
  Read Multiple Request, 224
  Read Request, 224
  in service-oriented architecture, 25

Signed Write Command, 225–226
state machines, 183–185
Write Command, 225
Write Request, 225
Attribute Protocol Layer
　asymmetric design at, 14–15
　security protection at, 16
Attribute types
　Find By Type Value Request/response, 222–223
　Find Information Request/response, 221–222
　fundamental, 192
　overview of, 190–191
　Unsupported Group Type error, 231
Attribute value(s)
　attribute permissions applying to, 194
　Characteristic Descriptor, 192
　Characteristic Type UUID, 192
　Find By Type Value Request/response, 222–223
　Handle Value Indication, 228
　Handle Value Notification, 227–228
　Invalid Attribute Value Length error, 230
　overview of, 191
　Prepare Write Request and Execute Write Request, 226–227
　Read Blob Request, 224
　Read By Type Request/response, 223
　Read Multiple Request, 224
　service UUIDs, 191
　units, 191
Attributes
　accessing, 196–197
　atomic operations and transactions, 197–198
　attribute handle, 189–190
　Attribute Protocol. *see* Attribute Protocol
　attribute type, 190–191
　attribute value, 191–193
　characteristics, 210–217
　grouping, 199
　overview of, 179
　peripheral design optimizing, 311–312
　permissions, 194–195
　structure of, 189
Attributes, background to
　data, data, everywhere. and, 180–181
　data and state, 181–182
　kinds of state, 182–183
　protocol proliferation is wrong, 180

services and profiles, 185–189
state machines, 183–185
Attributes, services
　combining services, 204–205
　extending services, 201–203
　including services, 209–210
　overview of, 199–201
　plug-and-play client applications, 207–208
　primary or secondary, 205–207
　reusing another service, 203–204
　service declaration, 208–209
Authentication
　attribute database permissions as, 194–195
　authorization vs., 195
　Bluetooth low energy and, 115
　in bonding process, 259
　central devices initiating bonding via, 292–293
　concept of, 241–242
　data channel, 30
　encrypted packet, 104
　Insufficient Authentication error, 229
　integrity via, 243
　pairing procedure, 250–251
　resolving signatures for, 225–226, 247
Authorization
　Insufficient Authorization error, 229
　security and, 242–243
Authorization permissions, attribute database, 195
Auto-connection establishment procedure, GAP, 267–268
Autonomy, service-oriented architecture, 24
Ax encryption blocks, encrypting payload data, 106

B

Bandwidth, classic Bluetooth and, 3
Basic Rate (BR), original Bluetooth, 3
Batteries
　lowering cost with button-cell, 5–6
　monitoring in connectionless model, 44
Behavior
　application layer services and, 37
　combining services, 204–205
　extending services, 201–203
　primary vs. secondary services and, 205–207
　profiles and, 37–38, 185
　reusing another service and, 203–204
　service characteristics and, 200–201
　services and, 34–36

BER (bit error rate), receiver sensitivity, 58
B-frame format, 32
Binary FSK (frequency-shift keying), digital modulation, 52
Bit error rate (BER), receiver sensitivity, 58
Bit errors
    CRC detecting odd numbers of, 84
    protection against, 16
Bit order
    access address and, 80–81
    packet structure and, 79
    preamble and, 79–80
Bit rate, optimizing for low power, 125–126
Bits, defined, 51
Block counter, encrypting payload data, 106–107
Bluetooth classic, fixed and connection-oriented channels, 170–171
Bluetooth classic vs. low energy
    compatibility with device types, 6
    connectionless model, 43–44
    overview of, 3–4
    power consumption, 8
    services and profiles, 185–189
Bluetooth low energy, overview
    concepts. *see* concepts
    design goals, 4, 7–8
    device types, 6
    low cost of, 4–5
    single-mode devices, 3–4
    terminology, 9–10
Bluetooth Qualification Administrator (BQA), 317
Bondable mode, GAP, 270
Bondable procedure, GAP, 270
Bonding
    central devices using, 292–293
    controlling connectability of peripherals, 301
    GAP defining device, 36
    long-term relationships and, 259
    modes and procedures for, 270
    optimizing peripherals for low power, 304–306
    profile security, 296–297
BQA (Bluetooth Qualification Administrator), 317
BR (Basic Rate), original Bluetooth, 3
BR/EDR Not Supported flag, advertising data, 274

Broadcast Flag, HCI data packets, 138–139
Broadcaster role, GAP, 261
Broadcasting data
    advertising state for, 71
    HCI, 148–153
    new wireless model for, 42–43
    overview of, 92–93
    Server Characteristic Configuration Descriptor for, 214–215
Broadcasting model
    active scanning, 152–153
    advertising, 148–150
    defined, 148
    passive scanning, 150–152
    peripherals that only broadcast, 299–300
Brute-force checking, private addresses, 261
Buffer sizes, HCI controller setup, 142–143
Bulk data USB packets, HCI, 134
Button-cell batteries
    concept of, 11–12
    lowering cost of Bluetooth low energy, 5–6
    short duration bursts of, 13
    single-mode devices designed for, 6
Bytes, packet structure, 79

C

Calibration, of controller in Direct Test Mode, 62
Categories, of qualification tests, 318–319
CCM (Counter with Cipher Block Chaining-Message Authentication Code Mode), 106
Cell phones
    dual-mode controllers for, 6
    marketing concept for, 19
    two-chip solutions on, 39–40
Central devices
    background of, 283
    bonding, 292–293
    building generic clients, 287–288
    changing services, 293–294
    connecting to devices, 285–286
    controlling connectability of peripherals, 301
    discoverability of peripherals, 283–285, 301
    implementing profiles, 294–297
    interacting with services, 288–292
    understanding, 286
Central role, GAP, 262
Changed services, central devices, 293–294
Channel identifiers, L2CAP, 172–173

Channel map
  HCI advertising, 150
  HCI connection management, 159–160
  Link Layer, 85
  Link Layer connection process, 97–98
Channel map, adaptive frequency hopping
  Link Layer connections, 94, 97–98, 111–112
  Link Layer robustness, 120–122
  overview of, 88–89
Channels
  Bluetooth classic using narrow, 55
  Bluetooth low energy using radio, 56
  HCI interface, 135
  L2CAP. *see* L2CAP (Logical Link Control and Adaptation Protocol)
  UART transport, 132–133
Channels, Link Layer
  adaptive frequency hopping, 88–89
  determining advertising vs. data packets, 76
  frequency hopping, 87
  overview of, 30–31, 84–85
  understanding, 84–87
Characteristic Aggregation Format Descriptor, 217
Characteristic Descriptors, attribute value, 192
Characteristic Extended Properties Descriptor, 214
Characteristic Presentation Format Descriptor, 215–217, 287
Characteristic Type UUID, 192
Characteristic User Description descriptor, 214
Characteristic Value Reliable Writes procedure, 237
Characteristic(s)
  application layer, 36–37
  central device discovery, 286
  central device interaction with services, 288–289
  combining services, 204–205
  declaration of, 211–213
  descriptors on, 214–217
  discovering with Read By Type Request, 223
  discovery and configuration of services, 258–259
  discovery on initial connection, 258
  exposing services to peripherals, 302–303
  extending services, 201–203
  GATT client-initiated procedures for, 235–238
  GATT discovery procedures for, 234–235
  grouping, 199
  optimizing peripheral attributes, 310–311
  overview of, 210–211
  peripheral devices, 302–303
  primary vs. secondary services, 205–207
  profiles discovering and using, 296
  reusing another service, 203–204
  services as grouping of, 37, 199–200
  value of, 213
Chips, defined, 51
Ciphertext, encryption text, 105
Classes, object-oriented programming, 199–200
Clear to send (CTS), 5-wire UART transport, 132
Client Characteristic Configuration Descriptor
  notifications and indications, 292
  overview of, 214
  profiles, 296
Client Preferred Connection Parameters characteristic, 285–286
Client-initiated procedures, GATT
  overview of, 235
  reading characteristic values, 235–236
  reading/writing characteristic descriptors, 238
  writing characteristic values, 236–238
Clients, building generic, 287–288
Client-server architecture
  asymmetric design of, 14–15
  attribute database and, 192–193
  attribute permissions, 194–195
  Attribute Protocol messages, 33
  concept of, 17–18
  data concept, 181–182
  as paradigm for Bluetooth low energy, 20–21
  profiles and services in, 186–189
  state-based model for, 17
Clock accuracy, Link Layer connection process, 98
CMAC algorithm, signing of data, 252
CMOS (Complimentary Metal on Silicon), 124–125
Command Complete event, HCI
  channel map update, 159
  command flow control, 139–140
  encryption, 145–146
  event packets, 137–138

Command Complete event, HCI (*continued*)
  reading device address, 141–142
  reading supported features, 143–144
  reading supported states, 144–145
  resetting controller to known state, 141
  setting random address, 147
  white lists, 147
Command flow control, HCI, 139–140
Command not understood reason code,
    command reject command, 174–175
Command packets, HCI, 135–137
Command reject command, LE signaling
    channel, 174–175
Command Status event
  enabling command flow control, 139–140
  encrypting data packets while connected,
    161–162
  HCI event packets, 138
  HCI feature exchange, 160
Commands
  Attribute Protocol, 218–219
  connection, 137
  controller state, 136
  Direct Test Mode, 65–68
  as exceptions to transaction rules, 197
  requesting specific action, 136
Company identifier, version information, 118
Compliance folder, testing and qualification,
    317–318
Complimentary Metal on Silicon (CMOS),
    124–125
Component subsystem product type, 315–316
Composability, service-oriented architecture,
    24
Concepts
  architectural paradigms, 20–25
    asymmetric design, 14–15
    button-cell batteries, 11–12
    client-server architecture, 17–18
    connectionless model, 19–20
    design for success, 15–16
    everything has state, 16–17
    memory is expensive, 13–14
    modular architecture, 18–19
    one billion is a small number, 19
    targeting new market segments, 11
    time is energy, 12–13
Confidentiality
  ensuring with encryption, 104
  security concept of, 243

CONNECT_REQ, advertising packet, 82
Connectable advertising state, peripherals,
    304–307
Connectable directed advertising, 149
Connectable modes, GAP
  direct-connectable, 266–267
  nonconnectable, 266
  overview of, 266
  undirected-connectable, 267
Connectable undirected advertising, 148
Connection events
  determining instant by counting, 112
  Link Layer connection process, 96–97
  optimizing for low power by subrating,
    128–130
  optimizing for low power with
    single-channel, 127–128
  sleep clock accuracy in connection process,
    98
Connection handle
  controlling connections with, 137
  HCI interface, 135
  labeling HCI data packets with, 138–139
  LE Connection Complete event, 155
Connection interval, optimizing peripherals,
    308–309
Connection management. *see* HCI connection
    management
Connection parameter update request
    command, LE signaling channel,
    175–177
Connection parameter updates, Link Layer,
    109–111
Connection Signature Resolving Key.
    see CSRK (Connection Signature
    Resolving Key)
Connection state, Link Layer state machine,
    73–74
Connectionless model
  achieving with L2CAP layer for.
    see L2CAP (Logical Link Control and
    Adaptation Protocol)
  new wireless model enabling, 43–44
  overview of, 19–20
Connection-oriented model
  channel identifiers for, 172
  connectionless model vs., 43–44
  Internet built around, 45
Connections
  controlling, 137
  establishing initial device, 258

Index

initiating from central devices, 285–286
peripheral devices, 301
reconnected, 260
Connections, creating at Link Layer
  access address, 95
  channel map, 97–98
  connection events, 96–97
  CRC initialization, 95
  initiating state for, 72
  overview of, 30–31
  sleep clock accuracy, 98
  transmit window, 95–96
  understanding, 93–94
Connections, initiating in HCI
  canceling, 156–157
  HCI initiating connections to devices, 156
  overview of, 153–154
  to white list, 154–155
Connections, managing Link Layer
  adaptive frequency hopping, 111–112
  connection parameter update, 109–111
  feature exchange, 118
  offline encryption, 130
  overview of, 109
  restarting encryption, 115–116
  starting encryption, 112–115
  terminate procedure, 118–119
  version exchange, 117–118
Connections, optimizing peripherals for low power
  bonding, 306
  connectable advertising, 306–307
  connected, 307–309
  directed advertising, 307
  discoverable advertising, 305
  overview of, 303–305
  stay connected or disconnect, 309–310
Consistency check, starting new project, 316–317
Continuation messages, LLID, 100–101
Control endpoint, USB interface in HCI, 134
Control points, Attribute Protocol
  central devices interacting with services, 289–290
  characteristics, 303
  defined, 183
  state machine, 183–185, 290–291
Controller
  configuring state of, 136
  device density design, 16

Direct Test Mode, 29–30
dual-mode, 6
HCI. *see* HCI (Host/Controller Interface)
Link Layer. *see* Link Layer
overview of, 27–28
Physical Layer. *see* Physical Layer
three-chip solution, 39–40
two-chip solution, 39–40
Controller subsystem product type, 315–316
Correlation of access address, 80–81
Cost
  design goal of low, 7–8
  designing Bluetooth low energy for low, 4–6
  memory is expensive concept, 13–14
  one billion is a small number concept, 19
Counter with Cipher Block Chaining-Message Authentication Code Mode (CCM), 106
CR2032 button-cell batteries, 11–12
CRC (cyclic redundancy check)
  3-Wire UARTs in HCI, 133
  bit errors and, 16
  calculating MIC, 107–109
  Link Layer connection process, 95
  Link Layer robustness with strong, 122–123
  overview of, 84
  packet structure, 30–31, 84
  Prepare/Execute Writes and, 198
  Prepare Write Request and, 227
  short range wireless standards, 8
  too weak to be security measure, 243
Create New Project page, bluetooth.org, 315
CSRK (Connection Signature Resolving Key)
  key distribution during pairing, 251
  long-term relationships, 259
  message authentication code, 226
  overview of, 247
  private addresses, 261
  signing of data, 252
CTS (clear to send), 5-wire UART transport, 132
Current time, peripherals that only broadcast, 300

D

Data
  packet structure, 30–31
  state vs., 181–182
  text packets transmitting, 63–64
  types in Bluetooth low energy devices, 180–181

Data access address, packet structure, 80–81
Data channels
   adaptive frequency hopping, 88–89
   frequency hopping over time, 87
   Link Layer and, 30–31
   placing, 84–87
Data flow control, HCI interface, 140
Data packets
   HCI interface, 138–139
   header contents, 82–83
   length field, 83
   overview of, 76
   starting encryption when connected, 161–162
Data packets, sending
   acknowledgement, 101
   example of, 101–104
   header, 99
   logical link identifier, 100–101
   more data, 101
   overview of, 98–99
   sequence numbers, 101
Data rates
   in classic Bluetooth vs. low energy, 3–4
   optimizing for low power, 125–126
   radio systems vs. application, 51
Data types, advertising, 273–276
DBm
   calculating range, 58–60
   measuring receiver sensitivity, 57–58
Debugging
   HCI version exchange, 160–161
   version information for, 117
Declaration, characteristic, 211–213
Declaration of Compliance (DoC), 313, 320–321
Description field, Characteristic Presentation Format Descriptor, 216–217
Descriptors, characteristic
   discovering all, 234–235
   discovery, central device, 286
   overview of, 214–217
   reading/writing, 238
Design
   asymmetric, 14–15
   compliance folder containing information on, 318
   goals, 7–8
   lowering cost, 4–6

service-oriented architecture goals, 21–25
   for success, 15–16
Development tool product type, 315–316
Device address
   HCI advertising parameters, 149–150
   HCI controller setup, 141–142
Device density, designing controller, 15–16
Device Name characteristic, GAP Service, 276, 284
Device Under Test. see DUT (Device Under Test)
Devices
   asymmetric design concept, 14–15
   Direct Test Mode requirements, 61–62
   finding, 90–92
   Generic Access Profile for, 36
   given tolerance of, 57
   initial connection to, 156, 258
   initial discovery procedure, 256–257
   new usage models for. see new usage models
   profiles describing two or more, 37–38
   time is energy concept, 12–13
   types of, 6
   types of data in Bluetooth low energy, 180–181
Digital modulation, 51–54
Digital radio, phase modulation in, 51
Digital television, 51
Direct advertising, 91–92
Direct Test Mode
   background of, 61–62
   controller architecture, 29–30
   hardware interface, 65–67
   transceiver testing, 62–65
   using HCI, 67–68
Direct-connectable mode, GAP, 266–267
Direct-connection establishment procedure, GAP, 269
Directed advertising, optimizing peripherals, 307
Discoverability
   advertising state used for, 71
   central device, 283–285
   Generic Access Profile defining device, 36
   initial discovery, 256–257
   modes, 264–265
   overview of, 263–264
   peripheral devices, 300–301
   procedures, 265–266
   in service-oriented architecture, 24–25

Discoverable advertising events, 82, 93
Discoverable advertising state, peripherals, 304–306
Discovery procedures, GATT, 232–235
DoC (Declaration of Compliance), 313, 320–321
Documentation, authorization via, 242–243
Dual-mode devices, 6
DUT (Device Under Test)
    Direct Test Mode, 61–62
    hardware interface, 65–67
    receiver tests, 64–65
    transceiver tests, 62
    transmitter tests for, 63–64
Duty cycle, short packets optimizing, 125
Dynamic refreshing, memory, 13–14

E

EDR (Enhanced Data Rate), Bluetooth version 2.0, 3
Encapsulation of services, 34
Encryption
    AES, 105–106
    authentication via, 242
    central device bonding using, 292–293
    data channel, 30
    ensuring confidentiality, 243
    HCI controller setup, 145–146
    HCI restarting, 163–164
    HCI starting, 161–162
    Insufficient Encryption error, 230
    Insufficient Encryption Key Size error, 230
    Link Layer restarting, 115–116
    Link Layer starting, 112–115
    Long-Term Key, 246
    lowering overhead with, 126
    message integrity check, 107–109
    offline, 130
    overview of, 104–105
    payload data, 106–107
    security design and, 16
    Short-Term Key, 246
Encryption Change event, HCI, 161, 163
Encryption engine, security, 244
Encryption Key Refresh Complete, HCI, 163–164
End product type, 315–316
Energy
    life of button-cell batteries, 12
    memory is expensive concept, 13–14
    time is, 12–13
Enhanced Data Rate (EDR), Bluetooth version 2.0, 3
Error Response, Attribute Protocol, 228–231
Errors
    bit, 16, 58, 84
    SDIO interface with low rates of, 135
    types of responses, 228–231
Ethernet, technologies increasing speeds of, 4
Event masks, HCI controller setup, 142
Event packets, HCI interface, 137–138
Events, Direct Test Mode, 65–68
Everything has state concept, 16–17
Exchange MTU procedure, GATT, 232
Exchange MTU Request and Response, Attribute Protocol, 221
Execute Write Request, Attribute Protocol
    characteristic descriptors procedure, 238
    characteristic values procedure, 236
    as exception to transaction rules, 198
    overview of, 226
    reliable writes procedure, 237
Extending services, 201–203
External state, 182

F

Features
    consistency check for new product, 316–317
    HCI connection management, 160
    HCI controller setup, 143–144
    Link Layer control, 118
    selecting for new product, 316
Filter policy, HCI, 150, 152
Filters
    Bluetooth low energy vs. classic, 29
    determining device discoverability, 257
Find By Type Value Request, Attribute Protocol, 222–223, 230, 233
Find Information Request, Attribute Protocol, 221–222, 230, 234–235
Find Requests, accessing attributes, 196
Finite state machines, Attribute Protocol, 184–185
Fixed channels, Bluetooth low energy supporting only, 171
Flags
    advertising data, 273–274
    HCI data packets, 138–139

Flags AD information
  advertising data, 273–274
  discoverable modes and, 264–265
  discoverable procedures and, 265–266
Flow control wires, 5-wire UART transport, 132
FM (frequency modulation) radio, analog, 51–52
Formal contracts, service-oriented architecture, 22
Format
  Bluetooth low energy requiring one frame, 32–33
  characteristic specification, 37–38
  test packet, 63
Format field
  Characteristic Aggregation Format Descriptor, 217
  Characteristic Presentation Format Descriptor, 215–216
Frame rate, 51
Frequency
  device tolerance and accuracy of, 57
  optimizing drift with short packets, 124–125
  peripherals that only broadcast, 300
  radio signal at Physical Layer, 28–29
Frequency bands
  agreements on allocation of, 51
  Bluetooth low energy using radio channels, 55–56
  overview of, 54
Frequency hopping
  adaptive. *see* adaptive frequency hopping
  Bluetooth classic using, 55
  data channels at Link Layer, 30
  defined, 9
  Link Layer connection process, 97–98
  overview of, 87
  spread spectrum radio regulations vs., 29
Frequency modulation (FM) radio, analog, 51–52
FSK (frequency-shift keying)
  Bluetooth low energy using GFSK, 54–55
  in digital modulation, 52
  MSK variant of, 53
  using whitener with, 77–79
Future-proof design, 18–19

G

GAP (Generic Access Profile)
  advertising data, 273–276
  attribute database including, 193
  background, 255–256
  bonding and pairing process, 252
  defined, 255
  establishing initial connection, 258
  exposing services to peripherals, 301–302
  generating private addresses, 106
  host architecture, 36
  initial discovery procedure, 256–257
  long-term relationships, 259
  private addresses, 260–261
  reconnections, 260
  roles, 261–262
  security modes, 270–273
  service characterization, 258–259
GAP (Generic Access Profile), modes and procedures
  bonding, 270
  broadcast mode and observation, 263
  connectability, 266–269
  discoverability, 263–266
  overview of, 262–263
GAP Service, 276–279, 284
Gateways
  client-server architecture, 17–18
  device interaction with Internet, 44–46
  modular service architecture and, 19
GATT (Generic Attribute Profile)
  characteristic discovery, 234–235
  client-initiated procedures, 235–239
  creation of, 179
  defining flat structure of attributes, 199
  discovering services, 232–233
  discovery procedures, 232
  ensuring future-proof design, 18
  forms of grouping, 200
  as GAP Service, 276–279
  host architecture, 34–36
  mapping ATT PDUs to, 239
  overview of, 231–232
Gaussian Frequency Shift Keying (GFSK), 28–29, 54–55
General advertising, 91, 93
General-connection establishment procedure, GAP, 268–269
General-discoverable mode, 256–257, 265–266
Generic Access Profile. *see* GAP (Generic Access Profile)
Generic Attribute Profile. *see* GATT (Generic Attribute Profile)

Index   333

Generic clients
    building for central devices, 287–288
    Characteristic Presentation Format
        Descriptor and, 215–217
    defined, 215
    enabling with GATT, 215
GFSK (Gaussian Frequency Shift Keying),
        28–29, 54–55
Global operations, 7–8, 54
Ground, 3-Wire UART transport, 132
Grouping
    Read By Group Type Request, 225
    services and characteristics, 199
    services using service declaration, 208–209
    Unsupported Group Type error, 231

H

Handle Value Indication, Attribute Protocol,
        228, 239
Handle Value Notification, Attribute
        Protocol, 227–228, 238
Hardware interface, Direct Test Mode, 65–67
Hash values, Identity Resolving Key, 246–247
HCI (Host/Controller Interface)
    active scanning, 152–153
    advertising, 148–150
    defined, 131
    Device Under Test requirements, 61
    Direct Test Mode using, 67–68
    initiating connections, 153–157
    overview of, 31
    passive scanning, 150–152
    segmentation and reassembly, 170
HCI connection management
    channel map update, 159–160
    connection update, 158
    feature exchange, 160
    initiating connections, 153–157
    restarting encryption, 163–164
    starting encryption, 161–163
    termination, 164–165
    version exchange, 160–161
HCI controller setup
    encrypting data, 145–146
    overview of, 140–141
    random numbers, 145
    reading buffer sizes, 142–143
    reading device address, 141–142
    reading supported features, 143–144

    reading supported states, 144–145
    resetting to known state, 141
    setting event masks, 142–143
    setting random address, 146–147
    white lists, 147–148
HCI Encrypt command, private addresses,
        261
HCI logical interface
    command flow control, 139–140
    command packets, 135–136
    data flow control, 140
    data packets, 138–139
    defined, 135
    event packets, 137
    HCI channels, 135
HCI physical interfaces
    3-Wire UART, 132–134
    overview of, 131
    SDIO, 134–135
    UART, 132
    USB, 134
Header
    data packet, 99
    framed packet, 133
    L2CAP packet, 173
    packet structure, 30–31, 81–83
Hop value, frequency hopping, 87
Host, enabling presence detection,
        41–42
Host architecture
    Attribute Protocol, 33–34
    attributes. see attributes
    Generic Access Profile. see GAP (Generic
        Access Profile)
    Generic Attribute Profile. see GATT
        (Generic Attribute Profile)
    L2CAP. see L2CAP (Logical Link Control
        and Adaptation Protocol)
    Logical Link Control and Adaptation
        Protocol, 32–33
    overview of, 32
    security. see security
    Security Manager, 33
    three-chip solution, 39–40
    two-chip solution, 39–40
Host subsystem product type, 315–316
Host/Controller Interface. see HCI
    (Host/Controller Interface)

I

ICS (Implementation Conformance Statements), 316–317
Identifiers, L2CAP channel, 171–172
Identity
   central devices discovering other device, 284
   Identity Resolving Key and, 246–247
Identity Resolving Key. *see* IRK (Identity Resolving Key)
IEEE 802.11, Bluetooth version 3.0, 3
IETF RFC 3610, encrypting payload data, 106
Immediate Alert Service, central devices, 290
Immutability, 200
Immutable encapsulation of services, 34
Imperial units, SI, 191
Implementation Conformance Statements (ICS), 316–317
Include attributes, services, 209–210
Include declaration, 233
Included services
   discovering, 233
   overview of, 209–210
   Read By Type Request searching for, 223
Indications
   accessing attributes, 196–197
   Attribute Protocol, 218–219
   central devices interacting with services, 291–292
   Client Characteristic Configuration Descriptor for, 214
   Handle Value Indication, 228
   optimizing peripheral attributes, 310–311
   server-initiated GATT procedure for, 239
   in service characterization, 259
Industrial, Scientific, and Medical (ISM) band. *see* 2.4GHz ISM band
Inheritance, enabling changes to interfaces, 200
Initial connection procedure, 258
Initial discovery procedure, GAP, 256–257
Initialization vector (IV), encryption, 114
Initiating connections
   from central devices, 285–286
   HCI, 153–157
Initiating state, Link Layer state machine, 73
Instant parameter, connection updates, 110–111
Insufficient Authentication error, 229

Insufficient Authorization error, 229
Insufficient Encryption error, 230
Insufficient Encryption Key Size error, 230
Insufficient Resources error, 231
Integrity, security concept of, 243
Interfaces, object-oriented programming, 199
Internal state, 182–185
International System of Units (SI), 191
Internet
   client-server architecture, 17–18
   gateways. *see* gateways
Interoperability
   Bluetooth classic/Bluetooth low energy, 6
   connection-oriented problems, 43–44
   profile/service architecture and, 185–189
Interpacket gap, optimizing for low power, 125
Invalid Attribute Value Length error, 230
Invalid CID in request reason code, 175
Invalid Handle error, attributes, 228–229
Invalid Offset error, 229
Invalid PDU error, 229
IP (Internet Protocol) license, 4–5
IPv6 (Internet Protocol), 46
IRK (Identity Resolving Key)
   key distribution during pairing, 251
   long-term relationships, 259
   overview of, 246–247
   saving during bonding for private addresses, 260–261
ISM (Industrial, Scientific, and Medical) band. *see* 2.4GHz ISM band
IV (initialization vector), encryption, 114

J

Just Works mode, TK value in, 245

K

Key distribution
   pairing procedure, 251
   security architecture, 15
   Security Manager protocol for, 33
Keys
   Connection Signature Resolving Key, 247
   encrypting text with, 105
   Identity Resolving Key, 246–247
   Long-Term Key, 246
   as shared secrets, 245
   Short-Term Key, 246
   Temporary Key, 245–246

L

L2CAP (Logical Link Control and Adaptation Protocol)
  background to, 169–171
  Bluetooth low energy using, 179–180
  channels, 171–172
  defined, 169
  host architecture and, 32–33
  LE signaling channel, 173–177
  optimizing peripherals for low power, 307–309
  packet structure, 172–173
  solving connection-oriented problems, 43–44
LANs (local area networks), 2.4GHz ISM band rules, 54
Latency, resolving low, 129–130
Layers
  defined, 9
  low power as design goal for, 7–8
LE Add Device To White List command, HCI, 147–148, 154–156
LE Advertising Report event, HCI, 152
LE Clear White List Size command, HCI, 147–148
LE Connection Complete event, HCI, 155–157
LE Connection Update command, HCI, 158
LE Connection Update Complete event, HCI, 158
LE Create Connection Cancel command, HCI, 157
LE Create Connection command, HCI, 154–157
LE Long Term Key Request event, 162–163
LE Rand command, HCI, 147
LE Read Advertising Channel Tx Power command, HCI, 150
LE Read Buffer Size command, HCI, 142–143
LE Read Channel Map command, HCI, 159
LE Read Remote Used Features command, HCI, 160
LE Read Remote Used Features Complete event, HCI, 160
LE Read Remote Version Information command, HCI, 160–161
LE Read Supported Features command, HCI, 143–144
LE Read Supported States command, HCI, 144–145
LE Read White List Size command, HCI, 147–148
LE Remove Device From White List command, HCI, 147–148
LE Set Advertising Data command, HCI, 150
LE Set Advertising Enable command, HCI, 150
LE Set Advertising Parameters command, HCI, 148–150
LE Set Host Channel Classification command, HCI, 159
LE Set Random Address command, HCI, 147
LE Set Scan Enable command, HCI, 152
LE Set Scan Parameters command, HCI, 150
LE Set Scan Response Data command, HCI, 150
LE signaling channel, L2CAP
  command reject command, 174–175
  connection parameter update request command, 175–177
  overview of, 173–174
LE Start Encryption command, 161–162
Leakage current, button-cell batteries, 12
Length field
  advertising data, 273
  packet structure, 30–31, 82–83
Licensing
  2.4GHz ISM band free of, 54
  Bluetooth low energy IP, 5
  Bluetooth low energy ISM band, 4–5
Limited-discoverable mode, devices
  discoverable procedures, 265–266
  initial discovery, 256
  overview of, 264–265
  peripherals, 300–301
Link budget, calculating range, 58–60
Link establishment mode, 3-Wire UART, 133
Link Layer
  advertising mode in, 41
  asymmetric design at, 14
  broadcasting, 92–93
  channels, 84–89
  controller architecture, 30–31
  creating connections, 93–98
  encryption, 104–109
  finding devices, 90–92
  function of, 69
  HCI. see HCI (Host/Controller Interface)
  low power as design goal for, 7
  managing connections, 109–119
  optimizing for low power. see optimization for low power
  packet structure, 79–84

Link Layer (*continued*)
  packets, 76–79
  robustness, 120–123
  sending data, 98–104
Link Layer state machine
  advertising, 71
  connection, 73–74
  multiple state machines, 74–75
  overview of, 69–70
  scanning, 72
  standby, 70–71
Link Loss Service, 288–289
Link Power Management, 134
LL_CHANNEL_MAP_REQ, 111–112
LL_CONNECTION_UPDATE_REQ, 109–111
LL_ENC_REQ, 112–113, 116
LL_ENC_RSP, 112–113
LL_FEATURE_REQ, 118
LL_FEATURE_RSP, 118
LL_PAUSE_ENC_REQ, 115
LL_PAUSE_ENC_RSP, 115–116
LL_START_ENC_REQ, 114
LL_START_RSP, 114–115
LL_TERMINATE_IND, 119
LLID (logical link identifier), data packet header, 100–101
Load balancing, client-server architecture, 21
Local area networks (LANs), 2.4GHz ISM band rules, 54
Local name advertising data type, 275
Logical interface. *see* HCI logical interface
Logical Link Control and Adaptation Protocol. *see* L2CAP (Logical Link Control and Adaptation Protocol)
Logical Link Control protocol, 180
Logical link identifier (LLID), data packet header, 100–101
Long-term relationships, bonding, 259
Loose coupling, service-oriented architecture, 22–23
Low power
  button-cell batteries for, 11–12
  as design goal, 7–8
  lowering cost of Bluetooth low energy with, 5–6
  optimizing for. *see* optimization for low power
Low power state mode, 3-Wire UART, 133
Lower-host controller interface, 31

LT (Lower Tester)
  Direct Test Mode, 61–62
  receiver tests, 64–65
  transceiver tests, 62
  transmitter tests, 64
LTK (Long-Term Key)
  key distribution during pairing, 251
  long-term relationships, 259
  overview of, 246
  private addresses, 261
  starting encryption for connections, 112–114

M

Man-in-the-middle attacks, 245–246, 249–250
Manufacturer-specific advertising data type, 276
Mapping
  ATT PDUs to GATT procedures, 239
  data broadcasting helping with, 42–43
  profiles to services, 37–38
Market segments
  one billion is a small number concept, 19
  targeted by Bluetooth low energy, 11
Master connection substate, 73–74
Masters
  asymmetric design concept of, 15
  defined, 9
  Link Layer connection process, 95–98
  multiple state machine restrictions, 74–75
Maximum transmission unit (MTU), Attribute Protocol, 221
Mbps (million bits per second), Bluetooth low energy transmission, 54–55
MD (more data) bit, 101–104
Memory
  Attribute Protocol requiring very little, 34
  cost of, 13–14
  Prepare Queue Full error and, 229–230
  single-chip solutions and, 39
Message authentication code, authentication signature, 226
Message integrity check. *see* MIC (message integrity check)
Metric units, SI, 191
MIC (message integrity check)
  AES calculating, 105
  encrypted packets including, 107–109
  encrypting payload data, 106–107
  Prepare/Execute Writes and, 198, 227

Million bits per second (Mbps), Bluetooth
low energy transmission, 54–55
Minimum-shift keying (MSK), 53, 55
Modems, technologies increasing speeds of, 4
Modes, GAP
  bonding, 270
  broadcast, 263
  connectable, 266–267
  discoverability, 263–265
  overview of, 262
  security levels and, 270–273
Modular architecture concept, 18–19
Modular service architecture, 18–19
Modulation
  analog, 49–51
  digital, 51–54
  overview of, 54–55
Modulation index
  Bluetooth low energy, 54–55
  digital modulation, 52–53
  radio signal, 29
More data (MD) bit, 101–104
MSK (minimum-shift keying), 53, 55
MTU (maximum transmission unit),
  Attribute Protocol, 221
Multiple state machines, 74–75
Multiplexing layer. *see* L2CAP (Logical Link
  Control and Adaptation Protocol)

N

Name, discovery of device, 257
NAT (network address translation), gateways,
  45
NESN (next expected sequence number), 99,
  101–104
Network address translation (NAT),
  gateways, 45
New usage models
  broadcasting data, 42–43
  connectionless model, 43–44
  gateways, 44–46
  presence detection, 41–42
Next expected sequence number (NESN), 99,
  101–104
Next expected sequence numbers, 101–104
NIST FIPS-197. *see* AES (Advanced
  Encryption System)
NIST Special Publication 800-38B, 247
Nokia, 5

Nonbondable mode, GAP, 270
Nonce, 106, 112–113
Nonconnectable advertising events, 82, 93
Nonconnectable mode, GAP, 266
Nonconnectable undirected advertising, 149
Nondiscoverable mode, 264
Nonresolvable private addresses, 278
Notifications
  accessing attributes, 196–197
  Attribute Protocol, 219
  central devices interacting with services,
    291–292
  Client Characteristic Configuration
    Descriptor for, 214
  as exception to transaction rules, 197
  Handle Value Notification, 227–228
  optimizing peripheral attributes, 310–311
  server-initiated GATT procedure for, 238
  in service characterization, 259
Null modem, UART configuration, 132
Num HCI Command Packets parameter,
  command flow control, 139–140

O

Object-oriented programming, 199
Objects, in object-oriented programming, 199
Observer role, GAP, 262
Offline encryption, 130
Offset, Invalid Offset error, 229
One billion is a small number concept, 19
Online resources, starting new project, 313
OOK (on-off keying), digital modulation,
  51–52
Optimization for low power
  acknowledgement scheme, 127
  high bit rate, 125–126
  low overhead, 126
  overview of, 123–124
  peripheral design for attributes, 311–312
  peripheral devices, 303–310
  short packets, 124–125
  single-channel connection events, 127–128
  subrating connection events, 128–130
Out Of Band algorithm, TK value in, 245
Overhead, optimizing for low power, 126

P

Packet Boundary Flag, HCI, 138–139
Packet counter, encrypting payload data, 106

Packet overhead, application data rate and, 51
Packet reporting event, Direct Test Mode, 67–68
Packet structure, Link Layer
　access address, 80–81
　bit order and bytes, 79–80
　CRC, 84
　header, 81–83
　length, 82–83
　overview of, 30–31, 76
　payload, 83–84
　preamble, 79–80
Packets
　advertising and data, 76
　as building block of Link Layer, 76
　CRC protecting against bit errors, 16
　initiating, 73
　optimizing with short, 124–125
　reducing memory requirements with small, 14
　restricting devices to short, 13
　structure of L2CAP, 172–173
　testing. see Direct Test Mode
　whitening, 77–79
Pairing
　authentication of link, 242, 250–251
　and bonding, 252
　central devices initiating bonding, 292–293
　exchange of information, 248–250
　key distribution, 251
　overview of, 248
　Security Manager protocol for, 33
　Short-Term Key for encrypting during, 246
　Temporary Key in, 245–246
Pairing Failed message, 249, 251
Pairing Request message, 249–250, 270
Pairing Response message, 249–250
PAL (Protocol Adaptation Layer), Bluetooth low energy, 169–170
PANs (personal area networks), 2.4GHz ISM band rules, 54
Parameters
　configuring advertising, 148–150
　HCI connection management by updating, 158
　HCI connections to white lists, 155
　HCI passive scanning, 150–152
　initiating connections from central devices, 285–286

Parity bit, UART, 132
Passive scanning
　central devices discovering devices with, 283–285
　HCI, 150–152
　Link Layer state machine, 72
　overview of, 72
　receiving broadcast data, 93
Passkey Entry mode, TK value, 245
Pathloss
　calculating link budget to determine range, 58–60
　central devices discovering devices, 284
Payload data
　3-Wire UARTs in HCI, 133
　AES encrypting, 105
　encrypting, 106–107
　L2CAP packet structure, 172–173
　packet structure, 83–84
PDUs, Attribute Protocol
　Invalid PDU error, 229
　mapping ATT PDUs to GATT procedures, 239
　overview of, 219–220
Peak current, button-cell batteries and, 12
Peripheral design
　background of, 299
　being connectable, 301
　being discoverable, 300–301
　broadcast only, 299–300
　characteristics, 302–303
　exposing services, 301–302
　optimizing attributes, 311–312
　optimizing for low power, 303–310
　security, 303
Peripheral Preferred Connection Parameters characteristic, GAP Service, 279
Peripheral Privacy Flag, GAP Service, 277–278
Peripheral role devices, GAP
　connectable modes, 266–269
　discoverability in, 263–264
　discoverability modes, 264–265
Permissions
　attribute database, 194–195
　Attribute Protocol, 34
　authorization via, 242–243
　profile security, 296
　security for peripherals, 303
Personal area networks (PANs), 2.4GHz ISM band rules, 54

Index

Phase modulation, 51
Physical bit rate, 51
Physical interfaces. *see* HCI physical interfaces
Physical Layer
  asymmetric design at, 14
  evolution of Bluetooth data rates, 3
  low power design goal for, 7
Physical Layer, controller
  analog modulation, 49–51
  architecture, 28–29
  background, 49
  digital modulation, 51–54
  frequency band, 54
  modulation, 54–55
  radio channels, 55–56
  range, 58–60
  receiver sensitivity, 57–58
  testing with Direct Test Mode, 29–30
  tolerance, 57
  transmit power, 56–57
Physical measurement, external state, 182
Piconet, 9
PIN (personal identification number), 104, 242, 244–245
Plan, test, 317
Plug-and-play client applications, 207–208
Power sensitivity, USB interface, 134
PRBS9 packet sequence, transmitter tests, 63–64
PRD (Qualification Program Reference Document), compliance, 320
Preamble, packet structure, 30–31, 79–80
Prepare Queue Full error, 229–230
Prepare Write Request, Attribute Protocol
  overview of, 198
  Prepare Queue Full error, 229–230
  reliable writes procedure, 237
  working with, 226–227
  writing characteristic descriptors procedure, 238
  writing characteristic values procedure, 236
Presence detection, new wireless model enabling, 41–42
Primary services
  defined, 37
  discovering all, 232–233
  discovering with service UUID, 233
  discovery, central device, 286
  Find By Type Value Request, 223
  grouping using service declaration, 208–209
  overview of, 35–36
  plug-and-play client applications, 207–208
  profile discovering for peer device, 295
  secondary vs., 205–207
Privacy
  creating with resolvable private addresses, 36
  Identity Resolving Key and, 246–247
  Peripheral Privacy Flag, 277–278
  primary goal of, 16
  security concept of, 243–244
Private addresses
  AES generating, 105–106
  complications of advertising using, 260
  defined, 260
  GAP connection procedures, 268–269
  for privacy, 16
  reconnection addresses as nonresolvable, 278
Procedures, GAP
  bonding, 270
  connectable, 267–269
  defined, 263
  discoverable, 265–266
  observation, 263
  types of, 263
Procedures, GATT
  characteristic discovery, 234–235
  client-initiated, 235–238
  Exchange MTU, 232
  mapping ATT PDUs to, 239
  overview of, 231–232
  server-initiated, 238–239
  service discovery, 232–233
Product information
  compliance folder contents, 318
  including in Declaration of Compliance, 320
Product types
  combining components, 321
  selecting features for new, 316
  selecting for Bluetooth low energy projects, 315–316
Profile subsystem product type, 315–316
Profile Tuning Suite (PTS) testers, qualification testing, 318
Profiles
  application layer, 37–38
  finding and using characteristics, 296
  finding services, 295
  generating test plan for, 317

Profiles (*continued*)
  modular service architecture for, 18–19
  security, 296–297
  selecting for new product, 316
  understanding, 294–295
Profile/service architecture
  in Bluetooth classic, 185–186
  in Bluetooth low energy, 186–189
Properties, characteristic, 211–214
Protocol Adaptation Layer (PAL), Bluetooth low energy, 169–170
Protocol messages, Attribute Protocol, 219–220
Protocol testers, qualification testing, 318
Protocols
  Bluetooth low energy, 179–180
  Bluetooth using Attribute Protocol. *see* Attribute Protocol
  memory burdened with multiple, 14
PTS (Profile Tuning Suite) testers, qualification testing, 318

Q

QDID (Qualified Design Identifier)
  combining components, 321
  declaring compliance, 320
  listing product, 321
  qualifying design, 319–320
Quadrature amplitude modulation, 51
Qualification program. *see* testing and qualification
Qualification Program Reference Document (PRD), compliance, 320

R

Race conditions, HCI, 157
Radio Band, 9–10
Radio channels
  overview of, 55–56
  starting receiver tests, 64
  starting transmitter tests, 63–64
Radio signals
  analog modulation and, 50–51
  controllers transmitting and receiving, 27
  enabling presence detection, 41–42
  high bit rate for low power, 125
  measuring path loss in, 58
  at Physical Layer, 28–29
  short range issues, 8
  widening of low energy, 29, 41

Radio-Frequency Identification (RFID) tags, 4
Random addresses
  HCI advertising parameters, 149–150
  HCI controller setup, 146–147
  Identity Resolving Key and, 246–247
  private addresses as, 260
Random numbers
  authentication during pairing, 250–251
  HCI controller setup, 145
  Long-Term Key using, 246
  Short-Term Key generated with, 246
  whiteners as, 77–79
Range, calculating, 58–60
Read BD_ADDR command, device address, 141–142
Read Blob Request, Attribute Protocol
  Attribute Not Long error, 230
  characteristic descriptors procedure, 238
  multiple characteristic values procedure, 235–236
  overview of, 224
Read Buffer Size command, HCI controller, 142–143
Read By Group Type Request, Attribute Protocol, 225, 230, 232–233
Read By Type Request, Attribute Protocol
  Attribute Not Found error, 230
  discovering all characteristics of service, 234
  discovering included services, 233
  multiple characteristic values procedure, 236
  overview of, 223
Read Characteristic Value by UUID procedure, central devices, 284
Read Multiple Request, Attribute Protocol, 224, 236
Read Not Permitted error, 229
Read only memory (ROM), single-chip solutions, 39
Read Request, Attribute Protocol
  accessing attributes, 196
  characteristic descriptors procedure, 238
  multiple characteristic values procedure, 235–236
  overview of, 224
Read Supported Features command, HCI controller, 143–144
Readable, access permission, 194

Readable and Writable, access permission, 194
Readable characteristics, 288
Readable state, 16–17
Reason codes, command reject command, 174–175
Receive data (RXD), UART/3-Wire UART transport, 132
Received signal strength (RSSI), central devices, 284
Receiver sensitivity, 57–58
Receiver test command, Direct Test Mode, 66, 68
Receivers
  in advertising state, 71
  analog modulation and, 49–51
  asymmetric design of, 14
  calculating range, 58–60
  time is energy concept of, 12–13
  transceiver tests, 62–65
  using whitener with FSK, 77–79
Reconnected connections, 260
Reconnection Address, GAP Service, 278
References
  combining services, 204–205
  extending services, 201–203
  reusing another service, 203–204
  services referencing other services, 200–201
Relationships
  accommodating between services, 35
  central device discovery of, 286
  central devices initiating bonding, 292–293
  creating permanently with Generic Access Profile, 36
  profile service, 37–38
Remapping process, adaptive frequency hopping, 88–89
Replay attack protection
  authentication via signatures, 242
  encrypted packets, 105
Request Not Supported error, 229
Request to send (RTS), 5-wire UART transport, 132
Requests
  Attribute Protocol, 218–219
  error responses to, 228–231
Reset command, Direct Test Mode, 66, 68
Reset command, HCI controller, 141
Resolvable private addresses, 260–261, 268–269
Restarting encryption, HCI connections, 163–164
Reusability
  behaviors limiting, 37
  of characteristics, 37–38
  in service-oriented architecture, 23
RF testers, qualification testing, 318
RFID (Radio-Frequency Identification) tags, 4
Robustness, Link Layer, 120–123
Roles
  GAP, 261–262
  profile, 294–295
ROM (read only memory), single-chip solutions, 39
RSSI (received signal strength), central devices, 284
RTS (request to send), 5-wire UART transport, 132
Rules
  2.4 GHz ISM band, 54
  access address, 81
  Attribute Protocol, 33–34
RXD (receive data), UART/3-Wire UART transport, 132

S

Scale, client-server architecture, 21
Scan Parameters Service, peripheral optimization, 309–310
SCAN_REQ, advertising packet, 82
SCAN_REQ packets, HCI active scanning, 152
SCAN_RSP, advertising packet, 82
SCAN_RSP packets, HCI active scanning, 152
Scannable undirected advertising, 149
Scanners
  asymmetric design of, 14–15
  enabling presence detection, 41–42
  initial discovery process, 256–257
  at Link Layer, 30–31
  receiving advertising events via, 91
Scanning state, Link Layer state machine, 72
Scatternets, 75
SDIO interface, HCI, 134–135
Secondary services
  defined, 37
  grouping using service declaration, 208–209
  including services, 209–210
  overview of, 35–36
  primary vs., 205–207

Secure Simple Pairing feature, 248–250
Security
  asymmetric design of, 15
  authentication, 241–242
  authorization, 242–243
  bonding, 252
  client-server gateway model of, 18
  confidentiality, 243
  Connection Signature Resolving Key, 247
  designing for success, 16
  encryption engine, 244
  Identity Resolving Key, 246–247
  integrity, 243
  Long-Term Key, 246
  overview of, 241
  pairing, 248–251
  peripheral devices, 303
  privacy, 243–244
  profile, 296–297
  shared secrets, 244–245
  Short-Term Key, 246
  signing of data, 252–253
  Temporary Key, 245–246
Security Manager
  Bluetooth low energy using, 179–180
  channel identifier for, 172
  host architecture, 33
  signing of data, 106
Segmentation, by multiplexing layers, 170
Selective-connection establishment procedure, GAP, 269
Sequence numbers (SNs), 101–104
Server Characteristic Configuration Descriptor, 214–215
Server-initiated procedures, GATT, 238–239
Service Changed characteristic, 294
Service data advertising data type, 276
Service solicitation advertising data type, 275
Service UUIDs
  discovering primary service, 233
  Include attributes, 209–210
  overview of, 191
  service advertising data types and, 274–275
  service declaration, 209
Service-oriented architecture
  abstraction, 23
  autonomy, 24
  composability, 24
  discoverability, 24–25
  formal contract, 22

  loose coupling, 22–23
  as paradigm for Bluetooth low energy, 21–22
  reusability, 23
  statelessness, 23–24
Services
  advertising data types for, 274
  application layer, 37
  central device changing, 293–294
  central device interaction with, 288–292
  central device's client remembering/caching between connections, 293–294
  combining, 204–205
  defining with profile roles, 294–295
  discovery at initial connect, 258
  extending, 201–203
  filtering advertising data based on, 257
  GATT characteristic discovery procedures for, 234–235
  GATT discovery procedures for, 232–233
  generating test plan for, 317
  Generic Attribute Profile and, 34–36
  grouping, 199, 208–209
  mapping profiles to, 37–38
  modular architecture for, 18–19
  optimizing peripheral attributes, 310–311
  peripheral design for exposing, 301–302
  plug-and-play client applications, 207–208
  primary or secondary, 205–206
  profiles discovering, 185–189, 295–296
  reusing, 203–204
  security for peripherals, 303
  selecting for new product, 316
Session based, connection-oriented model of Internet, 45
Session key diversifiers (SKD), 114
Session key (SK), 112–115
Shared secrets
  authentication via, 241–242
  in bonding process, 259
  Connection Signature Resolving Key, 247
  encrypting data packets while connected using, 161–162
  Identity Resolving Key, 246–247
  keys as, 245
  Long-Term Key, 246
  overview of, 244–245
  Security Manager for key distribution, 33
  Short-Term Key, 246
  Temporary Key, 245–246

Index

Shift register, 77
Short packets, for low power, 124–125
Short range wireless standards, 8
Short-Term Key (ST), 245–246
Short-wave radio, 51
SI (International System of Units), 191
SIG (Special Interest Group), Bluetooth
   testing and qualification requirements, 313–316
   UnPlugFest testing events, 15
Signaling channel, channel identifier for, 172
Signaling MTU exceeded reason code, command reject command, 175
SignCounter
   authentication signature, 226
   Connection Signature Resolving Key, 247
   signing of data, 252–253
Signed Write Command, Attribute Protocol, 225–226, 237–238
Signing of data
   AES, 105
   authentication via, 242
   Connection Signature Resolving Key, 247
   security and, 252–253
Silicon manufacturing processes, short packets optimizing, 124–125
Simultaneous LE And BR/EDR To Same Device Capable, 274
Single-channel connection events, 127–128
Single-chip solutions, stack split, 38–39
Single-mode devices, 6
SK (session key), 112–115
SKD (session key diversifiers), 114
Slave connection interval range, 275
Slave connection substate, 73–74
Slave latency
   connecting to devices, 285
   connection events and, 96–97, 129–130
   connection parameter update request and, 175–176
   connection update request, 111
   controlling in peripherals, 308–309
   defined, 129
   optimizing peripherals for low power, 308–309
Slaves
   in asymmetric design, 14–15
   connection parameter update request and, 109–111
   defined, 10

Link Layer connection process, 95–98
   multiple state machine restrictions, 74–75
Sleep clock accuracy, Link Layer connection process, 98
Sleep message, 3-Wire UARTs in HCI, 133–134
SLIP, framing packets in 3-Wire UART, 133
SNs (sequence numbers), 101–104
Spark-gap radios, 49–50, 51
Special Interest Group. see SIG (Special Interest Group), Bluetooth
Speeds, technology almost always increasing, 3–4
Spread spectrum radio regulations, 29
ST (Short-Term Key), 245–246
Stack splits architecture, 38–40
Standby state, Link Layer, 70–71
Start messages, LLID, 100
Starting encryption, HCI connection management, 161–163
Starting new project, qualification program, 313–316
State
   configuring controller, 136
   in connectionless model, 44
   in connection-oriented systems, 43–44
   data vs., 181–182
   HCI advertising filter policy, 150
   HCI controller setup, 141, 144–145
   kinds of, 182
   Link Layer. see Link Layer state machine
   optimizing peripherals for low power, 304–305
State machines
   Attribute Protocol, 183–185
   central devices interacting with services, 290–291
   Link Layer. see Link Layer state machine
   representing current internal state, 182
Statelessness
   of Attribute Protocol, 34
   in service-oriented architecture, 23–24
Stop bit, UART, 132
Subrated connection events, 128–130
Sub-version number, version information, 118
Symbols, 51

T

TCP connection, as session-based, 45
Temperature, button-cell batteries, 12
Temporary Key (TK), 245–246, 250

Termination
  error response resulting in request, 231
  HCI connections, 164–165
  Link Layer connections, 118–119
Test end command, Direct Test Mode, 66, 68
Test equipment product type, 315–316
Test Plan Generator (TPG) project, 313–315, 317
Test status event, Direct Test Mode, 67–68
Testing and qualification
  Bluetooth process for, 314
  combining components, 321
  consistency check, 316–317
  creating compliance folder, 317–318
  declaring compliance, 320
  generating test plan, 317
  listing, 321
  overview of, 313
  qualification testing, 318–319
  qualify your design, 319–320
  selecting features, 316
  standardizing. see Direct Test Mode
  starting project, 313–316
Testing information, compliance folder contents, 318
Text strings, associating with characteristics, 214
Third-party attackers, compromising integrity, 243
Three-chip solutions, stack split, 40
Three-way handshake, encryption for connections, 113, 115
Time is energy concept, 12–13
TK (Temporary Key), 245–246, 250
Toggle command, state machines, 184–185
Tolerance, 57
TPG (Test Plan Generator) project, 313–315, 317
Transactions, atomic operations and, 197–198
Transceiver testing, Direct Test Mode, 62–65
Transmit (TX) power level advertising data type, 275, 284
Transmit power, 56–57
Transmit window, Link Layer connections, 95–96, 110–111
Transmitter test command, Direct Test Mode, 66, 68
Transmitters
  in advertising state, 71
  analog modulation and, 49–51

asymmetric design of, 14
calculating range, 58–60
time is energy concept of, 12–13
transceiver tests, 62–65
Two-chip solutions, stack split, 39–40
TX (transmit) power level advertising data type, 275, 284
TXD (transmit data), UART/3-Wire UART transport, 132

U

UART (Universal Asynchronous Receiver Transmitter), HCI
  3-Wire, 132–134
  Direct Test Mode, 61, 65
  physical interface, 132
Undirected-connectable mode, GAP, 267
Unit UUIDs, 191
Units
  Characteristic Presentation Format Descriptor, 216–217
  generic client, 287
Unlikely Error response, 230
UnPlugFest testing events, 15
Unsupported Group Type error, 231
Updates
  adaptive frequency hopping, 111–112
  connection parameter, 109–111
Upper-host controller interface, 31
URLs, client-server architecture, 20–21
Usage models. see new usage models
USB physical interface, HCI, 134
UT (Upper Tester)
  Direct Test Mode, 61–62
  receiver tests, 64–65
  transceiver tests, 62
UUIDs (Universally Unique Identifiers)
  attribute types, 192
  Bluetooth Base, 190–191
  characteristic, 212–213, 236
  characteristics at application layer labeled with, 37–38
  discovering all primary services, 233
  Find Information Response and, 222
  generic clients and, 287
  identifying attribute type, 190
  service declaration, 209
  service UUIDs. see service UUIDs
  unit UUIDs, 191

V

Validated testers, qualification testing, 318
Value handle, characteristic, 212
Values, characteristic
  overview of, 213
  reading, 235–236
  writing, 236–238
Version exchange
  HCI connection management, 160–161
  Link Layer connections, 117–118

W

White lists
  auto-connection establishment procedure, 267–268
  connectability of peripherals, 301
  HCI advertising filter policy, 150
  HCI controller setup, 147–148
  HCI initiating connection to device(s) in, 154–156
  HCI passive scanning filter policy, 152
Whitening, 77–79, 81
Wibree technology, 5
Wi-Fi
  adaptive frequency hopping remapping, 88–89
  defined, 10
  Link Layer channels and, 84–85
  technologies increasing speeds of, 4
Window widening, 309
Wired infrastructure, problem of Internet design, 45
Wireless band, global operation design goals, 7–8
Woken message, 3-Wire UARTs in HCI, 134
Writable, access permission, 194
Writable characteristics, 288–289
Writable state, 17
Write Command, Attribute Protocol
  accessing attributes, 196
  Signed Write Command, 225–226
  writing without response procedure, 237–238
Write Request, Attribute Protocol
  accessing attributes, 196
  characteristic descriptors procedure, 238
  characteristic values procedure, 236
  overview of, 225

X

XML files
  characteristic specifications, 302–303
  generic clients and, 287–288

FREE Online Edition

Your purchase of **Bluetooth Low Energy** includes access to a free online edition for 45 days through the **Safari Books Online** subscription service. Nearly every Prentice Hall book is available online through **Safari Books Online**, along with thousands of books and videos from publishers such as Addison-Wesley Professional, Cisco Press, Exam Cram, IBM Press, O'Reilly Media, Que, Sams, and VMware Press.

Safari Books Online is a digital library providing searchable, on-demand access to thousands of technology, digital media, and professional development books and videos from leading publishers. With one monthly or yearly subscription price, you get unlimited access to learning tools and information on topics including mobile app and software development, tips and tricks on using your favorite gadgets, networking, project management, graphic design, and much more.

Activate your FREE Online Edition at informit.com/safarifree

STEP 1: Enter the coupon code: XXLNVFA.

STEP 2: New Safari users, complete the brief registration form. Safari subscribers, just log in.

If you have difficulty registering on Safari or accessing the online edition, please e-mail customer-service@safaribooksonline.com